First published in 2007
by The Squeeze Press

The Squeeze Press, Tigger Cottage,
Butleigh, Glastonbury, Somerset UK

© Vivian Linacre 2007
ISBN 1-978-906069-01-8

British Library
Cataloguing in Publication Data
Linacre, V., *The General Rule*
A Guide to Customary Weights and Measures

Printed and bound at
the Gomer Press, Llandysul, Wales

Set in 10pt Plantin

the
SQUEEZE
PRESS

THE GENERAL RULE

A Guide to Customary Weights and Measures

by

Vivian Linacre

DEDICATED

to the children of the Prime Minister and Chancellor
of the Exchequer whose fathers were in charge of the
government which had made the primary use of
imperial weights and measures a criminal offence
yet whose birth-weights were announced from
Downing Street and in the House of Commons in
Pounds & Ounces

and to two late lamented heroes

Steve Thoburn – the original 'Metric Martyr' – and
Norris McWhirter – athlete, scholar, patriot and friend

'STAND UP FOR THE FOOT'

Chris Jagger: from the album 'Atcha' – 1993

I like to measure my speed in miles per hour,
when I'm travelling down the road.

If I go to a bar I drink my beer in pints,
I use tons for a heavy load.

So don't give me kilometre, centimetre, millimetre,
parking metre, none of that crap.

I am a man who stands six feet tall and wears a ten-gallon hat

Stand up for the foot, it's feeling the pinch,
stand up for the foot and don't you give an inch.

I don't mind a litre of French wine
but don't tell me how to think.

If I want to buy me a pound of potatoes
or put a gallon of gas in my tank.

My feet are size nine, I can dance down the line
to gain me another thirty-two yards.

But tell me I've gone thirty metres
and I'm bound to kick your ass.

Stand up for the foot, it's feeling the pinch, stand up for the
foot and don't you give an inch.

You won't be reaching for a big ten-incher,
be a member of the 'mile high club',

Or go out looking for a pound of flesh
and having an ounce of luck.

So don't give me kilometre, centimetre, millimetre, parking
metre, none of that crap.

I am a man who stands six feet tall and wears a ten gallon hat.

Three brass yardsticks: a roller used for measuring cloth, between two County models, from respectively Bucks District and Lindsey part of Lincolnshire. Also a set of brass Troye weights from 50oz to 1oz; and merchant's weights from 7lb to 4oz. From the author's collection.

PREFACE

The clock and the calendar and the compass are all graduated in multiples or fractions of twelve – i.e. 4 x 3 – which makes them so flexible. 360 (12 x 30), the number of degrees in a circle, is divisible by 2,3,4,6,8 & 9 as well as 5 & 10. The basic rhythmical patterns in music are identified as 'six-eight time' (6/8), i.c. having 6 eighth-notes (quavers) to the bar; or 'common time' (4/4); or 'waltz time' (3/4), having 3 crotchets to the bar; or 'march time' (4/4 or 6/8), etc; always in factors of 2 and 3. Again, the binary system of numbers, the key to the internal coding of information in digital computers and hence to all modern technology, is based on the power of 2, increasing values on the scale of 1,2,4,8,16, etc. Hence the compatibility – scientific as well as practical – of our system of customary weights and measures, which is entirely composed of units factored on base 2 or 3.

For as Secretary of State John Quincy Adams said in his Report to Congress of 1821 (*you will find a full account in Appendix IV*): "Nature has no partialities for the number ten; and the attempt to shackle her freedom with it will forever prove abortive." Similarly, the great polymath Jacob Bronowski observed in '*The Ascent of Man*' (*see para. 12 of 'Epilogue'*): "So a crystal, like a pattern, must have a shape that could extend or repeat itself indefinitely. That is why the faces of a crystal can only have certain shapes; they could not have anything but the symmetries in the pattern. For example, the only rotations that are possible go twice or four times for a full turn, or three times or six times – not more. And not five times. You cannot make an assembly of atoms to make triangles which fit into space regularly five at a time."

So our duodecimal system has evolved because it is inherent in nature and therefore cannot be abolished. These customary measures were not invented, as the metric system was invented, but were discovered -- many thousands of years ago -- as demonstrated in *Megalithic Measures and Rhythms* (Floris Books, Edinburgh, 2006).

As *The General Rule* implies, there are exceptions. In music, the number five does figure in the major and minor thirds. The cross-section of DNA is a ten-fold rosette. The structure of viruses are based on icosahedral symmetry, which is 2-5-10-fold, and the recent discovery of five-fold quasicrystals, which are formed by supercooling liquid alloys extremely fast, also suggests that many liquids may be regarded as fluctuating pentagonal crystals. But none of this modern science invalidates the origins or evolution of customary measures.

The term 'customary' is used rather than 'imperial' because it has become an Anglo-American system. Just as there are more people on earth who speak American English rather than British English, so there are more people who use what Americans call 'English units' or the 'inch-pound' system than use what we still call the imperial system. Besides, 'Imperial' always was misleading, since it has nothing to do with Empire, but was merely a convenient label for the units that were standardized in 1826 (eliminating regional variations) once the industrial revolution had got under way. Yes, there are a few differences between UK and USA units (almost exclusively in fluid volume) but it's the same system; just as we speak the same language despite differences in spelling.

So our system more than matches metric in three of today's global industries – oil, computers and aviation – as well as in popular culture. Would anyone collect 'airkilometres' or compose a poem using any of those ugly polysyllables? Yet, for all its primordial origins, elemental nature, commercial predominance and colloquial ease, our customary system is under severe threat. The Napoleonic European Union is determined to make its use a criminal offence. Why don't they issue directives to metricate time and music too?

Hence the need for this book. Until a generation or two ago it would not have been necessary, for everybody in Britain and throughout the Commonwealth and the USA learnt customary measures along with their multiplication tables. Even after forty years of metric indoctrination, most young people today still give their height in feet and inches and weight in stones and pounds, while the world's two most popular children's authors, Roald Dahl and J K Rowling, have always used imperial measures. But the rest of the system is fading from British and even American minds. So something more is required than what we used to find on the back of exercise books: an introduction and companion to the whole traditional system, with an explanation of the relationships among the various sets of units, and of the historical and cultural background that gives the system so many extra dimensions.

I did produce a short guide for the British Weights and Measures Association in 2001, which generated more than enough enthusiasm and constructive criticism to warrant this definitive volume. The fascinating study of metrology and its evolution inevitably attracts mystical speculation which can tend to manipulate facts in order to suit esoteric theories. Some controversial material is included here, whatever its scientific validity, to sustain a sense of wonder and to illustrate the infinite ramifications of the subject. For who knows where ongoing archaeological and anthropological research into prehistoric metrology may yet lead us?

I must record my huge debt to John Strange for his extensive revisions and additional articles, to others who contributed new material or suggested improvements; to my BWMA colleagues, especially Michael Plumbe and Ralph Montagu; and to three authorities – Robin Heath, John Michell and John Neal – for their inspiration.

VIVIAN LINACRE
(President, British Weights and Measures Association)

CONTENTS

Introductory Articles 1

Standard Linear Measurements 13

Nautical Measurements 21

Astronomical Measures 24

Time 25

Miscellaneous Units and Measures of Length 29

Surveyors' Measures 35

Circular Measures 36

Weights and Mass 39

Avoirdupois Weights 40

Troy Weights 43

Maundy Pennies 44

Apothecaries' Weights 45

Measures of Area in Every Day Use 46

Gunter's Square Land Measure .47

Builders' Square Measures 47

Measures of Volume and Cubic Capacity 48

Fluid Measures 51

Weight of Water 53

In the Kitchen 53

Wine - Beer and Ale 54

Apothecaries' Fluid Measures 54

Medicine Measures 55

Paper Folding Sizes 56

Sizes of Bound Books 57

Paper Sizes 58

Music 60

Sound 62

Stress and Pressure 63

Temperature 64

Force 65

Work and Energy 67

Power 68

Speed 70

An Anthology 73

Megalithic Measures 79

Cosmic Numerology 83

Epilogue 97

Essential Knowledge 208

APPENDICES

I   Foreign equivalents of the foot and
Continental mile measurements 109

II   Quotations from Shakespeare 110

III   Parliamentary Reports from the Select Committee (1816)
and from the Commissioners (1819 and 1821) . . . 115

IV   Report from John Quincy Adams to the
United States Congress (1821) 119

V   The Gallon 122

VI   'Rules of Lawn Tennis' and Proof Spirit 125

VII   'The King's Girth'. 127

VIII   The Mystery of 42 137

IX   The Standards of Scotland 140

X   Binary Arithmetic 145

XI   A humorous adversary 153

XII   Questions and Answers:

        A   Weighing the Earth .155

        B   Days in Months 156

        C   Sunrise and Sunset .158

        D   Heat from the Sun 158

        E   Rainfall and Area 159

        F   Population and Area .160

G   The Big Bang 160

H   Mountains 161

J   More Mountains 162

K   Flying Bullets 163

L   360° Circles 163

M   Ordnance Survey 164

N   Earth's Interior 164

XIII   Wire Gauges 166

XIV   'Measuring America' 168

XV   The Beaufort Scale 172

XVI   Wine Levels 174

XVII   Metric Definitions of Imperial Units 175

XVIII   Coherent Systems of Units of Measurement 177

XIX   Hall-marking 180

XX   A short history of the Printer's Point 183

XXI   Threads in Engineering 186

Select Bibliography 189

The British Weights and Measures Association 191

Index of Sources 193

Index of Names and Places .195

General Index 197

A receipt of 1787 from an Exeter manufacturer of "all sorts of scale beams & money weights as by Act of Parliament", announcing quantities of gold and silver weighed and moved.

INTRODUCTORY ARTICLES

Any number is generally one of four types: *Decimal*, divisible into tenths, a system handed down by the Chinese and Egyptians; *Duodecimal*, divisible into twelfths, a system tracing from Western European pre-history and commonly used in the Roman Empire; *Binary*, divisible into halves, quarters, eighths, etc, originally Hindu, but fundamental now to computer mathematics; *Sexagesimal*, divisible into sixtieths, originally Babylonian and reflected today in the measurement of time and in geometry.

The last three of the four types of number – *Duodecimal*, *Binary* and *Sexagesimal* – are all compatible with the imperial system and with one another, as well as with computer mathematics which is the basis of modern technology. Several early civilizations, however, did use decimal counting: the Sumerian (Babylonian) adopted base 60, combining decimal and duodecimal, which gave the world units for the clock and the compass, while Phoenician merchants employed base 20, making a notch or *score* on a wooden stick when that number was reached. The Bible frequently refers to a score, and the French call 80 *quatre vingts*. Customary measures are very largely duodecimal, because 12 is far more easily factorized than 10, being divisible by 2, 3, 4 and 6 instead of only 2 and 5. It also lends itself to halving and doubling. Financial markets throughout the world quote and calculate yields and interest rates in fractions; from halves to quarters, to eighths, sixteenths and thirty-seconds – which is impossible in decimals. The mind naturally thinks in fractions. A child in Toulouse or Tokyo will say "I'm 8½" – never '8.5'! Rugby players wearing 11 and 14 on their backs are 'wing three-quarters', not 'wing 0.75s'. You agree to meet a friend in 'a quarter of an hour' – never in '0.25 of an hour'. Yet promoters of the metric system boast that it dispenses with the use of fractions!

With all systems of measure, certain basic definitions apply. Fundamental units are quantities whose scale of measurement is arbitrarily assigned and is independent of the scales of other quantities; all other quantities are derived units, subordinate to the original fundamental units. A unit of measurement is a precisely defined quantity, in terms of which the magnitudes of all other quantities of the same kind can be stated; whereas a standard of measurement is an object which, under specified conditions, defines, represents or records the magnitude of a unit.

The transatlantic dimension must be introduced at the outset, for the imperial system has become the Anglo-American system. Indeed, the main purpose of compulsory metrication is to deprive Britain of what the EU sees as our 'unfair competitive advantage' in sharing a common system of weights and measures with the world's superpower, to which UK governments respond: 'Quite right – it is iniquitous that Britain enjoys the colossal commercial and cultural benefit deriving from this common system of customary measures, which therefore must be abolished!'

There are a few, interesting differences between US units and ours, but the origins and system are the same; just as we share a common language despite minor differences in usage and spelling. Indeed, just as more people in the world speak American English than British English, so there are far more Americans using 'English' units of weight and measure than there are Britons using Imperial units.

Thus, in the British system, the dry capacity units and liquid units are the same, whereas in the American system they differ. In Britain the bushel was always used as a unit of capacity for both liquid and dry quantities, but in America is a dry measure only [see Fluid Measures]. Also, the UK liquid pint and dry pint each equal 568ml; about 20% larger than the US liquid pint of 473ml and slightly larger than the US dry pint of 550.61ml.

(William D Johnstone actually refers to "0.568 cubic decimeter", etc, but that would mean little even to fanatical UK metricists!)

To be precise, the UK pint = 568.26125ml and the US pint = 473.176473ml. Again, the common US *short ton* = 2,000 pounds and the British *long ton* = 2,240 pounds. Furthermore, in 1893 Congress decided that the metre should be 39.37in, making the inch very nearly 2.54000508cm. In 1897, the metre was legalized for trade in Britain; its value set at 39.370113in, making the inch very nearly 2.53999779cm. In 1959 the international inch was defined by the USA as 2.54cm. So 499,999 US inch of 1893 equalled 500,000 international inches; a difference of 1 inch in nearly 8 miles, which is negligible; although the US Coast and Geodetic Survey still use the 1893 inch. (New definitions for an international yard and pound were adopted universally on 1st January 1959 – see Appendix XVII)

However, the assumption that weights and measures must be standardized by fixed units is comparatively recent, deriving only from the concentration of power in the person of the sovereign at the expense of both the landed nobility and the liveried trade guilds – each of which throughout the Middle Ages defined and regulated their own customary measures to benefit their respective interests.

But the creation of a central government, emergence of a rapidly growing middle class, overseas possessions and a Royal Navy to protect them, all demanded a national code of weights and measures.

Furthermore, the Crown, struggling both constitutionally and financially to retain control over this modern state, had few sources of revenue apart from excise and other duties levied on goods – both domestically produced and imported – for which, however, there was a vastly increasing potential as manufacturing and overseas trade expanded rapidly. So fixed standards of weights and measures became imperative, in order to collect and account for the appropriate dues on a uniform basis across the country.

It is fascinating that previously the chief economic concern had been to ensure the integrity of the currency – always threatened by debasement and counterfeit – whereas standards of weight and measure were of minor importance and accordingly allowed to vary geographically and fluctuate periodically; but gradually priorities were reversed as stability of the means of exchange had become assured whilst stability of weights and measures was sorely needed. That preoccupation with the specification and regulation of a national system of weights and measures has intensified ever since, whilst concern for the credibility of money has declined! We should realize that this was not always so – that standardization of weights and measures is not necessarily sacrosanct and in earlier centuries was very properly regarded as of far less importance than safeguarding the value of money. Indeed, but for the fiscal plight of declining monarchies in Britain and Europe, and the profligacy of successive bureaucracies, protection of the pound sterling would have been given rather more attention than playing politics with the pound avoirdupois.

❖ ❖ ❖

from John Neal's monumental work

OPUS 2 - ALL DONE WITH MIRRORS

It is pointless to start at the chronological beginning of a study of metrology: to all intents and purposes there isn't one. Mathematical skills of some degree are evident among societies that were believed to have possessed only the most basic necessities for survival. At Ishango in Zaire during the 1950s, a bone was discovered, which was shown to be between 9,000 and 11,000 years old. It is squared off and perfectly straight, about nine inches long and resembles a handle with a sharp piece of quartz embedded in the hollow end that was perhaps used for scribing.

Scored upon its sides are a series of lines arranged in distinct groups. The archaeologist finder of the bone, Jean de Heinzelin, interprets the markings thus: in the first group the lines are 11, 13, 17 and 19 in number which he states are the prime numbers in ascending order. In the second, 3, 4, 6, 8, 10 and 5 then 5 and 7, which he believes suggests duplication by two coupled with decimal notation. The third group are 11, 21, 19 and 9 which he sees as 10 and 20 plus one and 10 and 20 minus one. He believes them to be tables of some description. Others maintain that it is simply some sort of tally or lunar count. Whichever is correct, the presence of the bone implies pre-historic numeracy.

Neither can one begin by trying to explain ancient metrology from the point of view of an evolutionary development; because the earliest known measures prove to be part of a completely developed system that on analysis is extremely sophisticated and identical in every respect to numerical systems used, the evidence suggests, universally. It would seem to be linked to the fully developed cultures that arose in the ancient world, also with no discernible evolutionary stages...

The modern search for a definition of the ancient measures begins in earnest with Isaac Newton. Before Newton, a Europe founded on Rome and [until the Reformation] dominated by the Church – after Newton, a world founded on science and dominated by industry and profit. Before Newton, there was little necessity to try to define ancient units of measurement as there was no application for them, nor would the implications of their geodetic relationships (*i.e. relating to measurement of the surface of the Earth*) have been understood, as no reliable measurement of the world existed.

Newton believed, by tradition, that the 'sacred cubit' of the Jews was a geodetic measure and, although he anxiously sought an accurate estimate of the Earth's dimensions to substantiate his laws of gravity, he also attempted to deduce this from analysis of ancient measurements. The modern objective search for these definitions can be said to have begun with Newton and continued to the present day.

That a circle has been measured since time immemorial as 1,296,000 arc seconds is truly remarkable, inasmuch as it is an extremely fine measure. (1,296,000 = 60 × 60 x 60 x 6) It is not unreasonable to conclude that the only reason for possessing a measure this fine would be for surveying on a geographic scale, and the fact that one second of arc of the meridian degree is 100 'Greek feet' is surely no coincidence.

But while the circle has been divided into 360° since pre-history, there is scant evidence that the degree was divided into 60 minutes before the last millennium BC, nor that the minute was divided into 60 seconds much before the Renaissance. That controversy continues.

Anyhow, the ancient Greeks could not have known that the earth was not a perfect sphere – i.e. that one second of arc, measured along a meridian, equals about 100.77 British feet at the equator, 101.02ft. at a latitude of 30° and 101.53ft. at a latitude of 60° – hence, perhaps, the variations in the 'Greek foot'?

Newton did not know that in 1664 one Richard Norwood, author of *A Seaman's Practice*, had taken observations of the sun at York (probably with a backstaff, since the sextant was invented by John Hadley in 1730) and another at Tower Hill in London, computing the degree to be 69.5 statute miles. But Newton's requirement of an accurate assessment of the length was finally satisfied by the efforts of a French astronomer, Jean-Luc Picard in 1671. His measurement of 69.1 English miles to the degree was not bettered for the next 150 years; it was a computation good enough for Newton to finalize his general theory of gravitation. The correct figure would have been between 69.114 and 69.115 miles. The average per degree is 69.055 miles, very close to the measure of 69.06 miles between 45° and 46°.

[*The British nautical mile is accurate at about 48° N – Picard's measurement was made in the English Channel at about 50°, so providing a basis for the British nautical mile* – see Nautical Measures.]

Probably the first reference to measures as statutes are those of the laws of Athelstan, who became king of Mercia on the death of his father Edward the Elder, in 934. The following year he became the first elected king of all England and was crowned at Kingston. His first tasks as sovereign were to establish the boundaries of his domain and institute just laws. [see Appendix VII.] The measures used by Athelstan, probably inherited by him from remote antiquity, and later defined by the proclamation of Edward I in 1305, were identical to those we use in the present day.

Philologists claim to have largely reconstructed the ancient vocabulary that indicates a vast commonality among ancient peoples...What has become apparent is that although their language may have mutated, their relatively uniform measurement system did not. It has been shown to have remained in continuous use from the prehistoric ages to the present day, throughout the lands populated by Eurasian peoples...beyond any boundaries that we believe they could possibly have crossed. Yet virtually nothing is known about this concrete evidence, while... scholars interminably discuss mere hypotheses concerning the origins of words and racial migrations!

The dissemination of the universal metrological system has been merely touched upon here. Just enough evidence has been presented to establish the general theory. So much is yet to be investigated: nothing has been mentioned of Angkor Wat, or even the Sumerian ziggurats, while the great pyramids and temples of China have received no attention. Tiahuanco in Bolivia is ripe for measurement, and a thousand other ancient monuments and forgotten cities from the Andes to the Gobi await the theodolite and yardstick to peer into their designs.

The ancient system is not primarily for quantifying…This may seem a contradiction in terms – non-quantitative mathematics – but a vast amount of mathematics is indeed non-quantitative. It is far more of a set of proportions than a means of calculating amounts. The system goes far beyond being based upon nature in that it is simply geodetically harmonious. The very means by which it divides by halving and increases by doubling is imitative of such natural processes as cell division.

Its counting bases more resemble those of the modern computer binary language: 16 bits, arranged in nibbles and bytes, a kilobyte being 1,024 bytes and 16,384 bits – all on base 4. Counting by tens expresses only the quantitative aspect of both ancient and modern systems, that are then applied to an underlying harmony of pure number, which has other means of dissemination than the decimal. [see Appendix X.]

How the binary system operates, may equally be regarded as a unitary system, for it is simply based on the number one and the absence of one, or zero. How very Pythagorean – 'All numbers are in one, and one is neither odd nor even'! Then, regarding metrology, this simple number is given a linear value, which is the English foot. Everything else may then be given its proportionate relationship and a means of comparison, and this has been seen to extend far beyond measures. It has spilled into ratios.

Yes, man *is* the measure of the world, and the cosmic model *can* be constructed from a Pythagorean triangle. The same light that shines forth from the crowded monuments of Egypt illumines stone circles on lonely hills in Scotland. Throughout Asia and the Americas, that same light is wherever you care to look – ruined, yet undiminished. Although the basic premise (prehistoric science) may be anathema to the prevailing orthodoxy, at least there are no institutions today that would actually brand the content as blasphemous or seditious.

Yet with the dawning of the year 2000, these ancient standards were declared illegal to use as trading standards in Britain, the country that had nurtured and propagated them!

Probably for the first time, this knowledge is revealed into a world that has forgotten it even exists. For it goes far beyond metrology, into the number system that governs nature or physics. It was not devised. Too many points of correspondence indicate that the system was a discovery.

It is no more an invention than Newton invented gravity or Einstein invented relativity – all they did was to devise mathematical models to explain natural phenomena – but this knowledge was understood in the ancient world with as great a clarity as those subjects are understood in the modern.

from Stephen Strauss's

THE SIZESAURUS

Merchants in mediaeval Danzig (Gdansk) would cloak inflation simply by changing measures. Thus the size of a purported 'pound's worth' of bread changed, depending on the cheapness of the wheat going into it, but the price stayed the same. One should perhaps pause and reflect about this other way of doing things for a moment…if some mediaevalist of a merchant tried to institute the elastic pound into his dealings today, his customers would think him not quaint but deeply immoral. Price can yo-yo up and down, and inflation can explode into hyperinflation, but in our minds pounds…are immutable. We imagine that to change them is in some way to change the laws of nature, and yet, only two hundred years ago, it was not clear at all that the only way to construct an economy was to keep the weights and measures fixed and fiddle with the money. Not clear at all.

But the most striking feature of metric's conversionary process has been the resistance of the old religions. Even though England had been asked to join at the very beginning, and the

American *philosophe* Thomas Jefferson proposed a decimalization of the English weight system in 1790 [he later rejected the metric system because it could not provide for conversion between time and distance measurement], an ecclesiastical war followed the birth of metric. Much of this undoubtedly reflected a continuing revulsion against the excesses of the French Revolution and its deification of reason over religion.

The ten-day-week metric calendar had not made provision for a Sabbath, and Christians in the United States in the nineteenth century found this proof that the system was essentially an abomination. [see (i) Time and (ii) Napoleon in An Anthology] We are all one weighing-and-measuring church, but that church's pews do not provide an altogether comfortable fit.

Metric is uniform, and its rigidity has foisted a uniform imperfection on the world. A gram is really too small to weigh anything but a single paper clip or a large raisin. Counting by tens is of course a rational way of measuring but....it often flaunts its precision in face of the lower level approximations of everyday life.

Asking for a pound of sugar seems more natural than requesting 500 grams of anything. Doubling or quartering of measures appeals to something quite primal in the human brain.* It ties into mental arithmetic... Metric eschews vulgar fractions and, in so doing, hides another truth... And that can be demonstrated by what everyone recognizes as the most common mistake in applying a metric measure. The easiest stumble is to be off by an order of magnitude (by misplacing the decimal point). Can anything be more odd? In the name of a universal precision we have allowed ourselves to be ten times more dramatically wrong than in the past. The vision of an absurdly inaccurate accuracy seems a fitting end to this little reflection on the two-hundred-year revolution in measurement and the ultimate installation of the metric god.

*Being born with two hands and two eyes, we innately conceive of a weight being held in one hand or the other, then half in each hand; which is why supposedly metric weights are forever halved and quartered, irrespective of the decimal units.

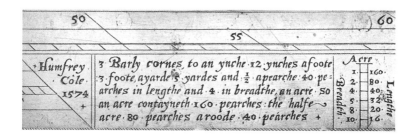

Detail of a 1574 ruler by Humfrey Cole with useful notes.

The Scots ell, used for measuring cloth, was standardized in 1661 as 37 inches. Ells were publicly displayed, for merchants to check their own measuring sticks. This one is mounted on a wall in Dunkeld, Perthshire.

Standard Linear Measure

1 foot = 12 inches
1 yard = 3 feet
1 rod, pole or perch = 5½ yards
1 chain = 4 rods = 22 yards
1 furlong = 10 chains = 40 rods = 220 yards
1 mile = 8 furlongs = 80 chains = 320 rods = 1,760 yards = 5,280ft

from MEASURE FOR MAN

*an article by the engineer and historical metrologist, Arthur Whillock,
published in The Dozenal Journal (1994)*

An appropriate unit of length is the basic requirement for any measuring system, so an anthropo-compatible range was an inevitable beginning. Names for components were transferred generically to comparable sizes as more definite means of determining them were found and simple relationships defined: some became more important than others, notably the Foot.

There were two forms of this. (a) The natural or ploughman's foot of twelve digits – about 9½ in – which in early days was half the Egyptian natural cubit or forearm-length of twenty-four digits; this was used by Celts and survives today in Wales and Scotland. (b) The visual foot, towards which derivations by many methods have tended, is the more practical by being the widest spacing that is comfortably perceived without head movement and therefore the most efficient to use for a hand-held measuring tool.

Visual feet [see Appendix I] range from 13.2in for the Indo-Aryan or Northern foot, described by the great Egyptologist Sir William Flinders Petrie (1853-1942) from a study of early buildings before an actual specimen was found, to 11.53in for the value of a Roman foot as used in southern England. The

Romans took their foot from an early Greek unit of sixteen digits, 11.8in, which they divided into twelve *pollices* (thumbs) for incorporating into their uncia system of weights and measures. Larger feet of 14in, found in such diverse areas as Poland, Italy and China, were probably half a *Braccia* or arm-length and favoured by cloth workers, an obvious measure for their purposes. The English measures common in the cloth trade until recently were the *nail* of $2^1/_4$ in – i.e. $^1/_{16}$th of a yard – and the *ell* of 45in or 20 *nails*.

William of Malmesbury stated that our yard was a distance given by the outstretched arm of Henry I (1106-1135) from nose to finger-tips, which must have been the 'Iron Ulna of our Lord the King' as mentioned in mediaeval documents. True or not, it was an appropriate way of fixing a length that all could appreciate – now affording an Aunt Sally for intolerant metricists who can only mock at simple things that they cannot understand.

There is no definite evidence on the origin of the English foot [see John Michell's references below: also in Megalithic Measures and Appendices I and VII], traces of which can be found in old buildings back to the 10th century. Both the Roman foot of 11.53in and the Greek common foot of 12.45in were used then for building work, so it has been suggested that our foot, which is known and used world-wide, was an average of the two, but units of length did not evolve that way. Many documents as far back as Saxon times give the length of a foot as *duodecim uncias pollices* (twelve thumb inches).

It was thus a true anthropometric measure that Edward I (1272-1307) legalized at three barleycorns to its inch. Flinders Petrie deduced that our foot has remained unchanged within 0.5% since then, being based on principles acceptable to the users. (How long would the metre survive without the aid of elaborate equipment and legal sanctions?) Of special interest, too, is the Greek *Olympic* foot, used for the layout of the Parthenon. This is a geodesic unit that divided a minute of arc on an assumed spherical Earth into 6,000 parts; i.e., 600 to a *Stade* being one tenth of a minute.

This is a controversial theory, but the parallel with our nautical mile is remarkable, as with the original measurement for the metre, which is nearly 0.2mm shorter than was intended. The Egyptians estimated the minute as 5,000 of their *Remen* units, so some authorities assume that the Greeks obtained their Olympic foot from them, but the sexagesimal format indicates a Sumerian origin. Accurate measurements like these would be possible on extensive river flood plains, with angles obtained from sun shadows or star sightings.

Human-sized measures are essential to allow correct perspectives toward our daily affairs. Just as a valid unit of length complies with visual acuity, so weight is judged by our muscular reactions to it. (Weight is a force, mass is not.) Piles of cobbles found at Neolithic sites were evidently the camp arsenal, ready to discourage prowling animals or repel unwanted visitors. These were no doubt selected for optimum range, accuracy and effect, with median weight somewhat less than one pound, appropriate for use by short stature people. (The 'war stones' collected by Pitt-Rivers from Polynesian natives were heavier.)

It is not too extravagant to suggest that an acceptable modern unit of weight should be a direct descendant of early throwing weapons. Cereal seeds, the units of settled living, by being readily available and appearing regularly in nearly the same sizes and weights, were widely used as reference standards for both.

To quote from Jacob Bronowski's '*The Ascent of Man*': "The largest single step in the ascent of man is the change from nomad to village agriculture. What made that possible? An act of will by men, surely; but with that, a strange and secret act of nature. In the burst of new vegetation at the end of the Ice Age c.10,000BC, a hybrid wheat appeared in the Middle East... wheat and water, they make civilization."

The original 'grain' weight was based on the wheat seed. Counts of barleycorns were also used for reference to the grain weight of the Hebrew *shekel* coin, whereas carob seeds became the medium for jewellers' weights (hence 'carats') for weighing gold and gems.

Edward I specified (*Statutum de Admensuratione*) 3 barleycorns for his English inch, 12 of which became the European standard foot; in contrast to the old Northern foot, which comprised 10in – followed by the longer Saxon foot and its yard, and subsequently to the legal value for the Rod, Pole or Perch measuring 5 Northern yards or $5^{1}/_{2}$ standard yards.

Hence the quarter chain, 40th furlong or 320th mile. This intermediate 'Saxon' regime provided for 1ft of 13.2in, 1 yard of 3ft, 1 rod of 5 yards, 1 furlong of 40 rods (600ft) and 1 mile of 8 furlongs (4,800ft), thus maintaining the $^{10}/_{11}$ ratio. As Whillock explained, Edward's new measure was essentially a 'Craftsman's foot', for use in building –"everything from cottages to cathedrals" – whereas the old Northern and Saxon foot, yard and rod were primarily for land measurement. The modern series of measurements based on the number 11 derive from the Anglo-Saxon system. Thus, the mile of 11 × 480ft, furlong of 11 × 60ft, chain of 11 × 6ft, rod/perch of 11 × $^{1}/_{2}$ ft, acre of 11 × 3,960sq.ft. (or 11 × 11 × 360sq.ft.).

Ian B Patten of Anchorage, Alaska, claimed in an influential article in The Kingdom Digest, Dallas, Texas: "Even the Anglo-Saxons were equally as astute in metrology as the Ancients. They had a dual system: linear units for straight lines – the digit

(0.72in), the fathom (72in), the inch and the foot – and those for land measurement, which bore a 11:10 ratio to the straight line units... Thus the link was 7.92in which equals $^{110}/_{10}{}^{\text{ths}}$ of the digit; while 10 links (79.2in) equal $^{11}/_{10}{}^{\text{ths}}$ of the fathom. The 100 links of the chain of 22 yards and the 1,000 links of the furlong ($^{1}/_{8}{}^{\text{th}}$ of a mile) make the sides of a rectangular acre of 43,560 sq.ft. And the 640 acres in a square mile was always so easily divisible into sections of 320, 160, 80, 40, 20, 10 and 5 acres. Furthermore, the Anglo-Saxon rod of 79.2in is exactly *double* the true measure of the metre at 39.6in, therefore equalling one ten-millionth part of one half of the Earth's circumference and hence coinciding with the 12-hour clock.

Again, this *true* metre of 39.6in is exactly $^{11}/_{10} \times$ the Imperial yard!

Hence, the metre fixed at 39.37in is an abomination, almost a quarter inch short – over nineteen feet short per kilometre. Even the earlier French system was more scientific than the metric; for the counterpart of the fathom was the *toise de Perou* standard, which had 864 *lignes* in its six *pieds*. 864, of course, is one hundredth of 86,400 which is the number of seconds in 24 hours.

So when metricists dismiss our duodecimal system as an antiquated hotch-potch they should look at the facts, which are that the Ancients and Anglo-Saxons knew far more about mathematics and metrology than a mob of revolutionary hot-heads a mere 200 years ago!"

He also declared: "To have any value, a land measurement must reflect a compatible time-distance relationship. The ancient Egyptians reckoned the Earth's rotational velocity at the equator is 1,521.75ft per second, and made their cubit 1,000th part of this: 1.52175ft or 18.261in. So every 4 seconds the Earth moved 4,000 cubits which is 1 nautical mile, equalling 1 minute of arc. Thus, there was total harmony between distance and time through the cubit as there is with the nautical mile today."

Such Egyptological aspects are warnings of far-fetched theorizing in this sphere! In fact, the Earth's equatorial radius is 20,925,646ft and its angular velocity is 72.92115 micro-radians per second; so the rotational speed at the equator is 20,925,646 × 72.92115 × (10 to the power minus 6) = 1,525.922ft per second. Still, an error of only $1/20^{th}$ of an inch by the ancient Egyptians in the astronomical measurement of the cubit wasn't a bad effort!

But there is no real evidence that the earth's polar flattening was even suspected prior to Newton. Besides, as John Michell has pointed out, there is no real relationship between the 79.2in rod and the Earth's circumference, which measures in *miles* exactly one tenth of 12 to the power 5 (see Cosmic Numerology). Again, the circumference of the Moon measures in feet exactly 12 to the power 7. This, indeed, is the principal standard of astronomical measurements. It is vital to understand that geodetic values are not derived from actual measurement but from an integral set of fractions: all of them fitting together by simple ratios. As Michell remarks: "The whole thing is a pre-existing number code, beautiful and subtle, which is surely the prototype of Creation." But, to come down to earth, it must be pointed out that the earth's (maximum) equatorial circumference is not quite 131,479,700 ft, half of which is only 9,960,584 Anglo-Saxon rods of 79.2". So this rod is about 4 parts per thousand too big for even the earth's greatest circumference, whereas the metre is only 2 parts per 10,000 too small.

In 1861 Sir John Herschel suggested a new "earth-commensurable" unit of linear measure which he called the *geometrical inch*; exactly one thousandth of an inch longer than the Imperial inch (1.001) and representing one 500 millionth part of the Earth's polar diameter. For he had discovered that the polar diameter measured 500,550,000 imperial inches. But we now know that the polar axis is about 500,531,700 inches, so 1.001 is not quite enough: 1.0010634 would be closer.

Herschel explained that by this tiny adjustment, a geometrical half-pint is exactly $^1/_{100}$th part of a geometrical cubic foot and a geometrical ounce is exactly the weight of $^1/_{1000}$th part of a geometrical cubic foot of distilled water. But 1 cubic foot actually contains 49.8306836723 pints – to be excessively precise – and if that number is multiplied by 1.0010634 *cubed*, we still get only 49.9898, not quite 50.00!

According to many Egyptologists, this same unit of 1.001in is the key to deciphering the Great Pyramid. For the Sacred Cubit of the Hebrews measured 25.027in just 25 geometric inches! So this polar diameter represents 20,000,000 and the polar radius 10,000,000 Sacred Cubits – this latter being the same number that the inventors of the metric system used (inaccurately) to define their basic unit in relation to the length of the polar *quadrant*. By the way, bearing in mind the importance that all ancient Middle Eastern civilizations attached to the number 7, it is perhaps relevant that 10,000,000 represents 10 to the power 7 [see also towards the end of Appendix VII].

Furthermore, considerable mediaeval evidence suggests that the former standard linear measures were indeed very slightly larger than the modern, the original standards having been lost in the House of Commons fire of 1834 – *not*, as frequently stated, the Great Fire of 1666!

Incidentally, taking the Great Pyramid inch as 1.001 British inch, then the polar semi-axis measures 250,266,000 B.in. or 250,016,000 GP.in. which = 10,000,600 sacred cubits. Now 10,000,000 sacred cubits = 3,949.653 miles, compared to the Earth's polar radius of 3,949.9 miles and the equatorial radius of about 3,963.19 miles. Close enough?

Eratosthenes (c.276-193BC) is generally credited with being the first to measure the Earth's circumference. When the Sun was vertically above Syene (modern Aswan), at mid-day at the summer solstice, it was 7°12 min. from the vertical at

Alexandria, where Eratosthenes was librarian. The two places are about 500 miles (5,000 *Stadia*) apart, but not quite on the same meridian; so he calculated the Earth's circumference to be 500 × 360 / 7.20 = 25,000 miles. This compares with the true distance of 24,900 miles – polar circumference 24,860 and equatorial 24,901 miles – not bad!

However, while Eratosthenes doubtless measured in degrees and minutes of degrees, it is unlikely that he could measure in seconds of minutes!

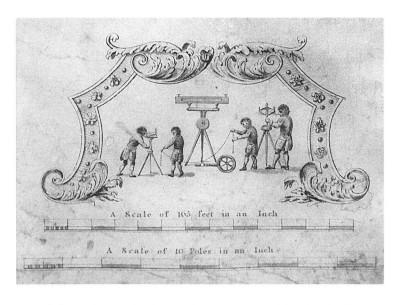

A decorative corner of an 18th century map of London, in the Westminster City Archives. Note the scales.

NAUTICAL MEASURES

1 fathom = 6 feet
1 log-line (dubious) = 450 feet = 75 fathoms
1 cable = 608 (say 600) feet = 100 fathoms
US cable (dubious) = 120 fathoms
10 cables = (approx) 1 nautical mile
1 nautical mile = 6,080 feet
    = $\frac{1}{60}$th of 1 minute of arc (1 degree of latitude)
    = $\frac{1}{60}$th of 69.091 (i.e. 1.1515) statute miles
1 knot = a speed of 1 nautical mile per hour

see also Appendix XV

A chain of 100ft was also used for marine surveying. The decimal series continues, since 100 fathoms = 1 cable and 10 cables = 1 nautical mile.

The fathom is the traditional unit for measuring depths at sea. It may derive from the length of an arm-spread of lead line; i.e. the length of one coil, but *fathom* derives from the Old English for 'embrace', because it measured from finger-tip to finger-tip of outstretched arms. For taking soundings, a line was attached to a lead weight which was swung in the direction the ship was moving, so that by letting go the line at the right moment and then hauling in the slack, a skilled leadsman could get an up-and-down sounding when he passed over the lead: an undemanding task given to an unfit seaman who was envied by the rest of the crew – hence the expression 'swinging the lead'? The lead-line was marked at 2 fathoms by 2 strips of leather, at 3 fathoms by 3 strips of leather, at 5 fathoms by a piece of white bunting and at 7 fathoms by a piece of red bunting, etc, and the corresponding depths were called 'marks', whereas unmarked depths were called 'deeps'.

Thus, if the depth was 2 fathoms the leadsman would call out 'by the mark twain' – hence the *nom de plume* adopted by Samuel

Clemens who had been a steamboat pilot on the Mississippi.

In olden times, ships carried a supply of logs for the galley fire. The speed of sailing was measured by heaving a log overboard, the time taken for it to pass the stern providing the basis of calculation against the known length of the ship. Readings were noted in a 'log-book'! This method was improved by knotting the rope at intervals of 47'3½" and running it out for 28 seconds – measured by a 28-second sand-glass – whereby it was calculated that a speed of '1 knot' = 6,080 ft. per hour, because $28 \times 6,080 / 3,600 = 47.29$ft. The more modern log consists of a rotator with fins on it, towed astern by means of a non-twisting log line, and can be used to measure distances as well as speeds.

A cable length differs from the cable attached to a ship's anchor, which is measured in *shackles* or *half-shackles*. Until about 1950, 1 shackle equalled 12½ fathoms = ⅛th of 1 cable; but since revised to 15 fathoms. The bower anchor cable of the ill-fated *HMS Hood* was 41 shackles of 12½ fathoms, a length of 3,075ft – over half a mile!

As explained in Epilogue 8, the Admiralty rounded off the nautical mile as 6,080ft = 1.1515 statute miles, but was superseded by the international nautical mile which is about 3ft 10½ in shorter at 1,852 metres. To convert nautical miles to land (statute) miles multiply by 1.15, and to convert statute to nautical miles multiply by 0.87. Since 1 statute mile = 5,280ft, 38 statute miles = 33 international nautical miles. 15 statute miles exceed 13 nautical miles by some 160ft, while 20 nautical miles exceed 23 statute miles by only 82ft ¾in.

In practice, the nautical mile is generally taken as 2,000 yards = 10 cable-lengths. The nautical mile is particularly handy, enabling the polar circumference of the Earth to be expressed as 21,600 miles (the number of minutes in a circle), corresponding to 24,860 statute miles. But today the international nautical mile is defined as 1,852 metres – about 3ft 10½ in shorter than the British standard.

Whereas the geographical mile measures one minute of longitude at the equator, giving a value of very nearly 6,087 ft ^{1}/$_4$ in, the nautical mile is measured along a meridian, its length varying with latitude. [see under Equivalents of 1 mile below] The latitude of Athens is about 38° N and the value of the nautical mile there is just over 6,06ft 4in. The value 6,080 ft, chosen by the Admiralty, is about right for 48°. At 50° N or S, it is 6,082 ft. The value 1,852 metres is appropriate in the region of Bordeaux.

Dividing the distance from the equator to the north pole, measured along a meridian, by 5,400, the average nautical mile is nearly 6,076ft 10in, ranging from just over 6,046ft 3in at the equator to 6,107ft 6in at the pole.

Tonnage, as applied to ships, has two quite distinct meanings. The Royal Navy uses 'full load displacement' tonnage the weight of the ship with all her equipment, represented by the weight of water she displaces. So the displacement of the mighty *Hood* was 41,000 tons. On the other hand, the mercantile marine the world over uses 'registered tonnage', which is the internal capacity of the ship measured in hundreds of cubic feet: so 1 registered ton = 100 cu.ft. Historically, a vessel's tonnage was the number of *tuns* or barrels she could carry; which explains why the French use the same word '*tonneau*' to mean both barrel and registered ton.

The size of a rope is its circumference in inches; the length of a piece of twine (easily found on board) needed to wind round the rope. The increasingly used alternative is the diameter of a rope measured in millimetres, but that requires a calliper (seldom found on board). Fortunately, the two measures are easily related; for the diameter in millimetres is roughly 8 times the circumference in inches [2.54 × 3.1416 or 25.4 /3.1416 = c.8.00]. So a 2^1/$_2$ in rope has a diameter of 20mm. Some small ropes are supplied in lengths of 20 fathoms and their size measured by weight: e.g. if we refer to a 2^1/$_2$ lb line we mean that 20 fathoms of it weighs 2^1/$_2$ lbs.

Astronomical measures

1 Light-second = 186,282.397 miles = 0.002003988805 astronomical units *(this is arithmetically precise but, like so many such calculations, unrealistically so!)*

1 Light-foot = the distance travelled by light in 1 nanosecond = 1 sec/10 to the power 9 (1 light-nanosecond = very nearly 11.802852677in – close to 1 foot!)

1 Astronomical unit (average between greatest and least distance of the Earth from the Sun) **= 92,955,807.2 miles = 499.0047835 light-seconds**

1 Light-year = 5,878,499,700,000 miles (the distance light travels in a year – i.e. 1 light second × the number of seconds in 1 year)

The speed of light is known exactly because distance is defined in terms of the speed of light (i.e. 299,792,458 miles/sec). But we can only know the light year with the precision with which we can measure the duration of the year, so any more precise representation would be absurd.

TIME

'Year' always means *tropical* year, as opposed to *sidereal* year, unless otherwise stated. They are differentiated by a phenomenon known as the 'precession of the equinoxes'. The sidereal year is the time taken for the Earth to complete its orbit about the Sun and the tropical year is the time that elapses between the Sun's crossing the equator at the beginning of Spring until its crossing the equator at the beginning of the following Spring.

The tropical year is 365.242193 days or 31,556,925.5 seconds, while the sidereal year is 365.25636 days or 31,558,149.5 seconds – and in this time the Earth rotates 366.25636 times, a day being lost by going round the Sun. So the sidereal day (1 rotation of the Earth) = 31,558,149.5 / 366.25636 or 86,164.1 seconds or 23 hours, 56 minutes and 4.1 seconds.

The synodic or lunar month is the period between two new Moons. Its average value is 29 days, 12 hours, 44 minutes and 2.9 seconds. So there are 12.36875 lunar months in a sidereal year. In that time, the Moon completes 13.36875 orbits about the Earth and thus the sidereal month, the time taken for the Moon to complete one orbit, is 27 days, 7 hours, 43 minutes and 11.5 seconds. Most ancient civilizations knew that if, theoretically, the Earth could travel on its own orbital diameter it would do so in 116¼ days (86,400 seconds in 24 hours) or 10,044,000 seconds. Now the constant velocity of light relates to the *mean* velocity of the Earth as 10044/1, while 116¼ days at the mean rate of 1,600,000 statute miles per day produce a 'circular' orbital diameter of 186,000,000 miles – therefore the constant velocity of light is 186,000 miles per second – compare with 186,282.397 miles (see above).

If the 1,600,000 miles of Earth's daily advance is divided by 86,400 seconds (24 hours) the resultant mean velocity per second is 18.5185185; the ratio of 186,000 to 18.5185185 being 10044/1. Robert W Casey described this ratio as "a unique thumb-print identifiable in the Milky Way as the development of intelligence...."

There is more. 502,200,000in or 7,926.136364 statute miles multiplied by *pi* (3.141592654) produces a circumference of 24,900.69177 miles. Dividing that by the 116¼ days relating to the Earth's diameter gives the '*pi*' year of 365.21046 days – very close indeed to the 365.2422-day solar or tropical year.

Now the scale of the Earth relative to its orbit is 1/23,466.6667 – equal to one revolution of our planet as an imaginary wheel, rotating at its mean orbital velocity, 23,466.6667 times in one solar year. Accordingly, the Earth's diameter of 7,926.136364 miles multiplied by 23,466.6667 produces the circular-orbit diameter of 186,000,000 miles.

By the way, astronomers seldom use light years. When dealing with our solar system they generally use astronomical units. Then, for the local group of galaxies they use *parsecs* and for the immense intergalactic distances they use *megaparsecs*. The 'parsec' of (648,000/*pi*) astronomical units (say 19,173,511,565,000 miles) = the number of seconds in an angle of 1 radian.

The Julian year of 365¼ days, used from 45BC until 1582, exceeds the tropical year by about 674.5 seconds. But the Earth is slowing down, days getting longer by about 0.006 seconds a year.

Incidentally, that covers only 1,626 years, because there was no year 0, although several readers of our 1st edition protested that there must have been; otherwise, without a zero between '– 1' and '+ 1', 2 years passed at the stroke of midnight! But others argue again that this is merely academic since the BC-AD calendar was constructed retrospectively. By 1582 the accumulated error was over 12 days, yet Pope Gregory XIII

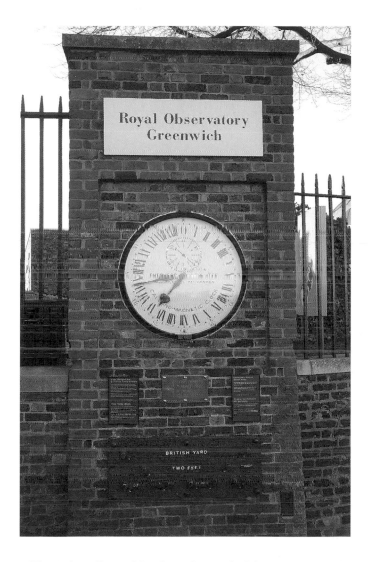

The clock at Greenwich, above the standard imperial yard and foot, close to the Greenwich meridian -- the zero line of longitude adopted in 1884 -- which runs through the old Royal Observatory and determines Greenwich Mean Time.

decided it was only 11 days; so Thursday 4 October 1582 was followed by Friday 15 October. But Britain (or, rather, England – Scotland having done so earlier) didn't change until 1752, by which time the gap was even bigger. The cry went up: "Give us back our 11 days!" – though there is no historical evidence of the alleged riots.

At the same time, the beginning of the year was put back from 25 March to 1 January. Hence the vestige of the Julian calendar and the reminder of the lost 11 days that still survives; the financial year ending on 5 April instead of on Lady Day – the 25 March quarter-day.

This Gregorian calendar, however, is a little more complicated. Year 'N' is a leap year if 'N' is divisible by 4 but not by 100, except that, if it is divisible by 100 then it is a leap year only if it is divisible by 400! So the year 2000 was a leap year but 1900 wasn't. The French Revolution led to the adoption of a decimal calendar on 5th October 1793. The year began on September 22 (1 vendémiaire) with twelve months each containing three 10-day weeks (décades) and ending with five *sans-culottides* and a further day in leap years to celebrate the Revolution. Each day was divided into 10 hours, each of 100 minutes, each of 100 seconds!

But on 7th April 1795 (the date on which the metric system was adopted) a further decree suspended its application to the clock. The Revolutionary calendar, however, lasted a further ten years until Napoleon I restored the traditional calendar on 1 January 1806, in exchange for Papal recognition.

Today, the only authorized metric unit of time is the second, and the only Système International *multiple* unit of time reached within a life-span is 1Gs, one gigasecond – i.e. a thousand million seconds, which is just under 32 years. Our allotted span of 'three score and ten years' becomes c.2.2Gs!

Miscellaneous Units and Measures of Length

Bed Sizes

Compact single	2'6" × 6'3"	Popular single	3'0" × 6'3"
Compact double	4'0" × 6.'3"	Popular double	4'6" × 6'3"
Queen size	5'0" × 6'6"	King size	6'0" × 6'6"

Grand Pianos

Baby	5'8"	Professional	6'0"
Drawing Room	6'4"	Parlour	6'8"
Half Concert	7'4"	Concert	8'11"

English Shoe Sizes

increase at $^{1}/_{3}^{rd}$ inch intervals:
size 8 = 11.333", 9 = 11.667", 10 = 12", 11 = 12.333" etc.
(depending on the style of shoe)

Biblical Lengths

finger = 1 digit (0.91in)   4 digits = 1palm (3.64in)
3 palms = 1 span (10.92in)   2 spans = 1cubit (21.85in)
6 cubits = 1 reed (10.92ft)   reed = 11ft
line = 80 cubits (but can also mean a 12th part of an inch!)
mile = 8 furlongs (then defined as Greek)

These are the orthodox Talmudic measures. But the 'primitive' (Zereth) measures include, interestingly enough, a span of 12.59in and cubit of 2 spans or 25.19in. That was probably the cubit that Jacob took into Egypt, where the use was made of the 'Egyptian Royal' cubit and the 'Olympic' cubit. So the Talmudist units emerged, having absorbed Babylonian influences, following the Jewish return from captivity.

Also, these measures from the New Testament:
4 cubits = 1 fathom, 100 fathoms = 1 stade and 8 stades = 1 mile

COTTON MEASURES

1 thread = 54 inches   1 skein = 80 threads
1 hank = 7 skeins   1 spindle = 18 hanks

Therefore the number of threads (inches) in 1 spindle = 544,320 (a number divisible by every number from 1 to 10), equal to 8 miles 1,040 yards.

NUMBERS USED IN TRADE

1 long (baker's) dozen = 13*   1 gross = 144
1 long gross = 156   1 score = 20
1 common hundred = 100   1 long or great hundred = 120

*Bakers used to give 13 to the dozen to preclude the risk of prosecution for short measure

FILM

35mm = 1.375"

Standard paper-backed film was 2³/₄" wide, but Edison simply cut it in half for use in ciné film, giving a width of 1.375". The German firm Ernst Leitz decided to make miniature still cameras using this narrow film, previously thought too narrow for still photography, and so the Leica camera was born. The film was accordingly dubbed 35mm, which equals 1.378". 70mm (2³/₄") film is used by large cinemas globally.

SHOTGUN BORES

The gauge of a shotgun barrel was originally determined by the number of identically-sized lead balls that have the same diameter as the bore, and weigh one pound. In other words, 12 lead balls, each with the diameter of a 12 gauge bore, weighs one pound – 20 equally sized balls, whose total weight is one pound, would identify the 20 gauge. The '410' is the only exception, this referring to a bore of 0.410".

PRINTER'S TYPE SIZES

The point is used for measuring font sizes in printing, athough the 'Anglo-Saxon' and European points are different. It originates from the age of metal type but has now been modified for computer typesetting [see Appendix XX].

American / British:
1 point = 0.01383 inch (approx $1/72^{nd}$ inch)
12 points = 1 Pica
72 points = 0.9962 inch

Continental Europe:
1 point (Didot) = 0.0148 inch
12 points (Didot) = 1 Cicero
72 points (Didot) = 1.0638 inch (1 French Royal inch)

Postscript (the most common computer type system):
1 point = $1/72^{nd}$ inch *exactly*

EQUIVALENTS OF 1 INCH

25,400.05 microns *(microns relate to the 1893 American inch)*

4,000 silversmiths' points
1,000 mils (mils are used for gauging wire and firearm bores)

64 *ounces* (to measure thickness of shoe leathers:
also a term in Troy weights)

48 hairsbreadths

12 *lines* (used in cloth measure and in printing)

3 barleycorns   1.3333 digits ($4/3$)   0.3333 palm ($1/3$)
0.25 hand ($1/4$)   0.08333 foot ($1/12$)   0.055556 cubit ($1/18$)
0.027778 yard ($1/36$)   0.005051 rod ($1/198$)

Many of these equivalents are not accurate because expressing fractions by decimals often requires use of 'repeaters'.

DERIVING FROM THE HUMAN BODY

1 digit	=	width of middle finger ($^3/_4$ inch)
1 inch	=	..........thumb across knuckle
1 nail	=	$2^1/_4$ inches
		(from middle joint to tip of middle finger)
1 ell	=	20 nails = 45 inches
		(from one shoulder across chest to finger-tip of outstretched arm – used for measuring cloth)
1 palm	=	3 inches
1 hand	=	4 inches
		(width of palm plus adjacent thumb joint)
1 shaftment	=	9 digits
		(hand + length of extended thumb – 6.00/6.55in)
1 span	=	9 inches (tips of outstretched thumb to little finger)
1 cubit	=	18 inches (from elbow to outstretched finger tips
1 fathom	=	6 feet (from tip to tip of fingers of outstretched arms)

EQUIVALENTS OF 1 FOOT
see also Appendix I

12,000 mils	144 lines
4 palms	3 hands
0.66667 cubit ($^2/_3$)	0.3333 yard ($^1/_3$)
0.060606 rod ($^2/_{33}$)	0.001515 furlong ($^1/_{660}$)

EQUIVALENTS OF 1 MILE

5,280 feet	3,520 cubits
1,760 yards	320 rods
8 furlongs	0.868979 meridian mile
0.868421 British nautical mile	0.867419 geographical mile

(1 geographical mile = 1.152845 statute miles and 1 statute mile = 63,360in, providing the scale of Ordnance Survey maps)

The *meridian* mile was established by international agreement in 1954 to represent 1 minute ($\frac{1}{60}$th of 1 degree) of the Earth's meridian, but has been generally replaced by the nautical mile, as also has the *geographic* mile (formerly known as the *admiralty mile*) which represented $\frac{1}{21,600}$th part of the Earth's equatorial circumference. The degree ($\frac{1}{360}$th of the Earth's equatorial circumference) is taken to equal 69.1707 statute miles; but 1° of latitude equals 68.708 miles at the equator and 69.403 miles at the poles, as determined by the International Astronomical Union Ellipsoid of 1964, while 1° of longitude equals 69.1707 miles at the equator diminishing to zero at the poles. [see Nautical Measures]

The young George Washington holding the surveyor's essential tools. In his left hand, Gunter's chain, in his right a circumferentor or primitive theodolite. The 22 yards of Gunter's chain can be found in the length of a cricket pitch and in the dimensions of every city block in the United States.

Surveyors' Measures

1 link = 7.92in
1 pole = 25 links = 16^1/$_2$ ft = 198in
1 chain = 100 links (792in) = 4 poles = 66ft
1 furlong = 10 chains = 1,000 links = 40 poles = 660ft
1 mile = 8 furlongs = 80 chains = 320 poles = 8,000 links

from a review in *The Times* 17 July 2002 of *Measuring America* by Andro Linklater

"It was a Professor of Astronomy at Gresham's College in London, Edmund Gunter, who invented his surveyor's chain in 1607. With 100 links divided into 4 equal rods or perches, it succeeded in combining traditional measurements with decimals. Consequently, 'Gunter's chain' became the standard rule and 'has since dictated the dimensions of our background in everything from the length of a cricket pitch to a New York city block'.

But Gunter's chain had a more profound impact; it enabled the Anglo-Saxon world to transform land into property, especially in the United States which, in the course of its independence, was to acquire 23 billion acres of additional territory, half to be sold off to individual ownership. Before this transformation could occur, however, the land had to be divided into transferable parcels. And so began the gigantic, rolling, 'gunterized' survey of the American continent, leaving this huge nation divided into a grid of small squares from the Appalachian Mountains to the Pacific Ocean." [see also Appendix XIV]

Decimal conversions	*multiply by*
To change feet into miles (nautical)	0.0001645
feet into miles (land)	0.0001894
feet per second into m.p.h.	0.6818
miles (nautical) into miles (land)	1.1515
miles (land) into miles (nautical)	0.8684
m.p.h. into feet per minute	88

A Victorian armillary sphere, showing the geometrical relationship between the rotating heavens and the earth. Despite the best efforts of metric campaigners, circles worldwide, whether heavenly or nautical, are still reckoned as consisting of 360 degrees.

CIRCULAR MEASURES

1 minute = 60 seconds
1 degree = 60 minutes
1 sextant = 60 degrees
1 quadrant = 90 degrees
4 quadrants = 1 circle
1 grade = 0.01 quadrant
        (where the right angle is divided into 100 grades)
1 circle = 360 degrees = 32 points (of the compass) = 4
quadrants = 400 grades = $2 \times {}^{22}/_7$ radians
1 radian = 57.2958 degrees
1 circle = $2 \times$ 'pi' $\times 1$ radian = $2 \times 3.14159 \times 57.2958$ = 360 degrees

The length of the arc of a circle that subtends an angle of 1 radian at the centre is equal to the radius of the circle – i.e. If a radian is subtended by an arc equal to the radius of a circle, the angle will measure 57 degrees 17 minutes and 48.8 seconds say 57.3 degrees. *Pi* can be approximated by dividing one radian by the cubit measured in inches – thus: 57.3 / 18.24 = 1.3141.

Conversely: 57.3 / 3.141 = 18.24. Now, 18.24 could never be as much a numerical constant as *pi*, because there are many varieties of cubit, but a clear relationship does exist between 18.24 and the 360 degrees of a circle, since 360 / (18.24 × 2) = 9.868, whose square root is 3.141. Again, 18.24 × 180 = 3,283, whose square root is 57.3. Evidently, the inch, the cubit and *Pi* are fundamentally connected.

The circle is also divided into the 32 points of the compass and the 12 signs of the zodiac – creating angles of respectively $11^1/_4°$ and 30°.

Part set of Troye weights, from 8lb to 1oz, Lanark, 1618, which has been completed by addition of later 1 stone and 1lb weights (courtesy of Museum of Edinburgh).

WEIGHT AND MASS

edited from Colin R Chapman's

HOW HEAVY, HOW MUCH AND HOW LONG

Lochin Publishing Society 1995 – ISBN 1 873686-09-9

In many cases we should be identifying mass, not weight. Strictly, the weight of an item is the force exerted on it by gravity, while mass is the amount of matter in that item. If, for example, the weight of an item is measured by a spring-balance at the bottom of a mine and then on the top of a mountain, it will weigh less on the mountain top as the gravitational attraction there is less, even though its mass has not altered. On the Moon, where the gravitational force is a sixth of that on Earth, the item will weigh six times lighter and a spring balance will indicate this.

In practice, for goods being weighed in a particular market on a beam balance or a steelyard [a type of balance having its fulcrum off-centre, enabling heavy goods to be weighed with relatively small weights: e.g. a 7lb weight at 24in. from the fulcrum counter-balancing a 28lb item at 6in. on the other side of the fulcrum], the difference between mass and weight is of no consequence; two items with equal masses having equal weights under identical conditions.

So the veracity of any iron and brass weights is important and therefore national and local standards, which were the subject of a number of Acts of Parliament, were used to check the weights used by tradesmen. [see also *Force*]

AVOIRDUPOIS WEIGHTS

more properly avoir de pois – goods of weight

1 ounce (oz) = 16 drams
1 pound (lb) = 16 ounces
1 stone = 14 lb
1 quarter = 2 stones
1 hundredweight (cwt) = 8 stone = 112 pounds
1 ton = 20 hundredweight = 160 stones = 2,240 pounds
1 short ton (US) = 2,000 pounds (20 centals of 100 pounds)

'Pound' derives from the Latin *pondus* meaning weight, and the abbreviation 'lb' from the Latin *libra* meaning pound. 'Ounce', like 'inch', comes from *uncia* meaning a twelfth part.

The old Saxon (or 'Tower') pound contained 5,400 grains. King Edward I had a 'merchant's pound' of 15 × 450 grains = 6,750 grains.

Then in 1340, Edward III's *haber de pois* weights (there's a set in Winchester Museum) were 16 Florentine ounces @ 437 grains (close to the Roman ounce of 438 grains) = 6,992 grains, which Henry VIII in 1527 finally rounded up to 7,000 grains (i.e. the ancient Egyptian Sep weight!).

So the avoirdupois pound was based on 16 ounces, equating to 7,000 grains; the avoirdupois ounce equalling (as it still does) $437\frac{1}{2}$ grains. The ease of 16 for division, especially in the wool trade, led to this system's becoming standard. Specific containers – sacks, barrels, casks, pockets, tubs, chests, baskets, drums, etc – became identified as standard measures for specific commodities.

But wool was always a special case, because of the wealth its trade created and the tax revenue levied on it, based on weight measured in trones (scales) using 7 and 14lb lead or bronze weights. The stone of 14lb is still widely used in the wool, fish and vegetable trades, as well as for weighing people.

The term *tonne* is increasingly used, frequently in error. It is the French word for the metric ton (megagram) and therefore a unit of *mass*, whose value is only about 35lb 6oz short of the British ton; i.e. very much closer to our ton than to the US *short ton* with which it is often confused. Moreover, contrary to common usage, it has nothing to do with the 'registered ton', in regard to trading vessels around the world, which is a unit of *volume* – equal to 100 cubic feet.

A French book of reference, *Le Marin de Commerce*, defines the tonneau as 100 cubic feet ('pieds'). So to describe a ship as the media do constantly – as (say) "7,000-tonne" is meaningless, except with reference to its deadweight, which is hardly ever what was intended! A ship's 'gross tonnage' measures her total internal volume whereas her (net) registered tonnage measures the volume used for cargo or passengers, while the 'displacement tonnage' (for warships) measures the weight of the vessel in terms of tons of seawater displaced when she has all her equipment on board.

The late Roland Smith, a Hastings butcher, with his steelyard, still in everyday use in 2001. The meat is suspended from the end hook and the beam from the other hook, while the counterpoise is moved along the arm until the beam is horizontal, when the position of the counterpoise relative to a scale engraved or notched on the long arm of the beam gives the weight reading. The great advantage is that no loose weights are required.

TROY WEIGHTS

formerly used for precious metals and gems

24 grains = 1 pennyweight (dwt)
20 dwts = 1 ounce
12 oz = 1 lb

So the Troy pound = 5,760 grains, while 240 dwts = 1lb – just as, in happier times, 240d = £1

The Troy pound equals 144/175 (i.e. 5,760/7000 grains) of the pound avoirdupois but the Troy ounce equals 192/175 of the ounce avoirdupois. An ancient English measure of weight (though named after the French city of Troyes), the Troy pound equated to the monetary pound sterling that contained 240 pennies – a penny originally containing a pennyworth of almost pure silver ($^{37}/_{40}$ silver with $^{3}/_{40}$ copper) which had the same weight as 24 grain]s of barleycorn.

Compare 24 grains with the 27.34375 grains of 1 drachm ($^{1}/_{16}$th oz avoirdupois). An Act of 1853 declared the Troy ounce of 120 *carats* to be the standard for the sale of bullion. The Troy pound was abolished in 1878, but the Troy ounce still survived (as 1.097oz avoirdupois) in the Weights and Measures Act of 1985. The smallest metal Troy weight is normally 10 grains ($^{1}/_{48}$th of 1 ounce), so for even smaller units – from 1 to 5 grains – tiny lengths of wire are used, called 'riders', twisted so that they can be slipped onto the beam of a balance. Hence the expression, 'adding a rider', meaning to make a minor qualification or adjustment.

Maundy Pennies

Maundy pennies of sterling silver are still minted but they weigh only seven and three-elevenths grains, comprising six and eight elevenths grains of silver plus six elevenths grains of base metal (copper) – because sterling silver is $^{37}/_{40}{}^{\text{ths}}$ pure silver.

But while that is the correct specification, the standard for sterling silver still remains 11oz 2dwts in the Troy pound (with 18dwts of alloy) as set by King Henry II over 800 years ago. There are, however, two standards for gold and two for silver: 22 or 18 carats of pure gold in every ounce containing 24 carats (although under the new regime since 1975 18 carat is stamped '750'); and for silver 11oz 10dwt or 11oz 2dwt – although the higher is very seldom used. (The old 'copper' penny weighed just $^{1}/_{3}{}^{\text{rd}}$ oz avoirdupois.)

The modern weight of the carat, formerly equivalent to 259.19564 milligrams, is standardized at 200mg: the word deriving from the Greek *keration* meaning the fruit of the carob, whose weight is remarkably constant from one specimen to another.

APOTHECARIES' WEIGHTS

formerly used by pharmacists

1 scruple = 20 grains	1 drachm = 3 scruples
1 ounce = 8 drachms	1 pound = 12 ounces

This drachm of 60 grains must not be confused with the drachm or dram of 27.34376 grains belonging to the avoirdupois system. The Apothecaries' and Troy pound and ounce alike equalled in weight to 5,760 and 480 grains of barleycorn. But the Apothecaries' system excluded pennyweights and properly excluded pounds too; dividing instead into drachms and scruples. It was abolished in Britain in 1971. Troy and Apothecaries' Weights remain full of interest, however, and references to them are frequently found.

Frequent references are likewise found to other ancient maritime and mercantile measures, which were universal and essential in their day, such as in this typical shipping report from 1821: *Arrived Greenock, 15 January, The Royal Charlotte from Lisbon, with 183 chests and 334 half-chests oranges, 4 pipes and 4 hogsheads white wine, ¹/₄ cask port wine, 144 quintals of salt and 10 dozen mats.* A quintal (or cental) was a weight of 100lb, often used for fish, while a mat(t) was a weight of 80lb, used for sugar and spices.

Unlike people today who argue that the metrically educated cannot deal in traditional measures, merchants of several different civilizations 4,000 years ago, trading from India to North Africa, had no difficulty in converting from any one system of weights to another. Thus, an ingot of the decorative stone lapis lazuli weighing 3lb was equivalent to 100 Dilmun shekels from the Indus valley, 160 Mesopotamian shekels from Babylonia or 175 Syrian shekels. Or, 15 Dilmun shekels weighed the same as 16 Egyptian *dbn*. (By sheer coincidence, apparently, the Ugarit and Syrian shekel and Egyptian *kdt* all weighed the same – about ¹/₃ʳᵈ of an ounce, or 4 dozen to the pound.) Easy!

Measures of Area in Every Day Use

1 square foot = 144 square inches
1 square yard = 9 square feet
1 square rod/pole/perch = $30^{1}/_{4}$ square yards
1 rood = 1 furlong × 1 rod = 40 square rods
    = 1,210 square yards = $^{1}/_{4}$ acre

1 acre = 4,840 square yards = 1 furlong × 1 chain
    = 1 furlong × 4 rods/poles/perches = 4 rods × 40 rods
    = 1 chain × 10 chains

1 square mile = 640 acres [i.e. 1,760 × 1,760 = 4,840 × 640]

The acre was originally the extent of land that a horse or ox could plough in one day. *('linacre' may derive from 'field of flax', or could it be an abbreviation of 'linear acre'?)*. It is 40 rods or 1 furlong ('furrow long') by 4 rods or 1 chain in width, divided into 72 furrows, 11 inches apart: i.e. 792 inches, 100 links or 22 yards.

1 cubic foot = 1,728 cubic inches
1 cubic yard = 27 cubic feet

Gunter's
Square Land Measure

see also Surveyors' Measures and Appendix XIV

1 square link = 62.7264 square inches
1 square rod/pole = 625 square links = 30¼ square yards
1 square chain = 16 square poles or 10,000 square links
1 rood = 40 square rods   1 acre = 4 roods = 10 square
chains 1 square mile = 640 acres
The *acre-inch* of 3,630 cu.ft and *acre-foot* of 43,560 cu.ft are
sometimes used in relation to irrigation and flood control

Builders' Square Measures

1 square of flooring = 100 square feet
1 rod of brickwork = 272½ square feet
1 bay of slating = 500 square feet
1 'yard' of land = 30 acres
1 'hide' of land = 120 acres (the 'long hundred')

Measures of Volume and Cubic Capacity

1 cubic foot = 1,728 cubic inches
1 cubic yard = 27 cu.ft
1 cu.ft. of water = 6.25 gallons weighing 1,000 ounces apx)
1 standard gallon = 277.274 cu.in
1 standard bushel = 8 gallons = 2,218.192 cu.in
1 boll (Scots) = 3 strikes = 6 bushels = 24 pecks = 48 gallons

1 cubic foot = 6.228835459 gallons and 1 cubic foot of water at 62°F weighs very nearly 997.68 ounces. [see Appendix V]

The standard gallon was defined in the 1824 Act; and revised to 277.421 cubic inches in 1932: but since 1878 a standard Imperial gallon has been defined as the capacity of 10 *avoirdupois* pounds of distilled water at 62°F and 30 inches of mercury barometric pressure.

Oil is measured in barrels rather than drums because, according to legend, when Edwin Drake first discovered 'rock oil' in Titusville, Pennsylvania, in 1859, the only suitable containers readily to hand were whisky barrels with their traditional capacity of 42 gallons. *[see also 'Fluid Measures' and 'Wine' and Appendix VIII]*

The timber industry happily deals in acres/hectares, cubic feet/metres, hoppus feet (21% short of a true cubic foot, to measure a log with its sides squared off), quarter girth (approximate area of a square fitted within the circumference of a tree), cants (squared logs), cords (piles of timber roughly 8ft x 4ft x 4ft high) and basal area (volume of trees in metres squared).

Stack of 11 brass standard fluid measures, stamped 'Board of Trade 1883', from 8 gallons (1 bushel) to 1/4 gill at the top. On the left a case of weights in grains, from 4,000 to 10 grains. On the shelf behind, the original measures for the pint, quart and gallon, stamped 'Geo.IV 1824', the year that introduced standard imperial units. Displayed in the old Jewel Tower, Palace of Westminster.

50

Upper: Set of 6 liquid capacity measures, from 1 gallon to 1 gill, Leith, c.1816 (courtesy of Museum of Edinburgh). Lower: Set of brass nesting cup 'trone' weights, Glasgow, early 19th century, from 4lb to 1/8th oz (courtesy of Clydebank Museum).

Fluid Measures

1 gill = 5 fluid ounces
1 pint = 4 gills = 8 tots /noggins = 20 fluid ounces
1 quart = 2 pints 1 pottle = 4 pints
1 gallon = 4 quarts = 8 pints
1 peck = 2 gallons 1 bushel = 4 pecks = 8 gallons

'Pint' appears to derive from the Latin *pingo* meaning 'I paint', from the mark that used to be painted on a vessel to indicate the pint measure. Bushels and pecks were customary measures, particularly for dry goods such as grain, salt and flour, even fish and coal, while less familiar measures were also employed. Barrels and hogsheads, with pints and gallons, were used as measures of dry capacity as well as indicating volumes of fluids - although the absolute size of a barrel depended very much on the nature of its contents.

Like the barrel, the gallon had an erratic pedigree, being defined in 1290 by its capacity for 8 pounds of wheat, then in 1706 as containing 231 cubic inches – which remains the USA standard. (The Pilgrim Fathers sailed in 1620 but the mass migrations took place much later.) The royal decree that led to the creation of dry and liquid measures had been King Henry VII's in the late 15th century, declaring that "eight pounds do make a gallon of wine and 8 gallons do make a bushel". In the USA they still do (see below). [see Appendix V]

So the Americans have their own fluid ounce, which is one sixteenth of a US pint, making the US fluid ounce fractionally bigger than the British one, perhaps compensating for the fact that the American gallon is smaller – roughly $5/6$ths of a British gallon. When an American refers to a 'quart' of spirits he means a bottle of 32 liquid ounces or 33.3 British fluid ounces, and by a 'fifth' (of a gallon) he means a bottle holding exactly the same as our bottle of Scotch – i.e. $1/6$th of an imperial gallon.

1 US liquid ounce = 1.04084 fl.oz = 1.80469 cu.in.

1 UK fl.oz. = 1.73387 cu.in.

1 cu.ft. = 6.22884 UK gallons = 7.48052 US gallons
    = 0.178108 US barrels

1 US gallon = 0.832674 UK gallons
    = 0.0238095 US barrels = 0.133681 cu.ft.

12 US gallons = (say) 10 UK gallons

1 UK gallon = 0.028594 US barrels = 277.419cu.in.
    = 0.160544cu.ft.

1 US barrel = 42 US gallons = 34.9723 imperial gallons
    = 5.61458 cu.ft.

1 US pint/gallon = 0.832674 UK pint/gallon

1 UK pint/gallon = 1.20095 US pint/gallon

Originally, the British gallon weighed 8lb, whether liquid or 'corn' (for dry goods); the pint bearing a 1:1 relationship to the pound. But when the Imperial system was established in 1826 (implementing the Act of 1824), a new capacity measure was introduced, the Imperial gallon. This replaced the Queen Ann Wine Gallon for liquid measure and the Winchester Bushel for dry measure, which were both smaller than the new unit. It is these two measures that are still used in the USA for their respective liquid and dry capacity measures. At the time of independence they simply adopted the British standards already in the country and have not altered them since.

It is interesting that the American quarter is just over 946ml whereas the old French pint is usually reckoned to have been about 930ml.

To recapitulate: gallons (Imperial and US), barrels, litres and cubic metres are all measures of volume and therefore can be converted easily between metric and customary units. So can tons (long, short and metric), which are all measures of weight (see below). But conversion from barrels to tons/tonnes is more difficult, depending (e.g.) on specific gravity, particularly in the case of oil. Specific gravity varies according to the American Petroleum Institute's tables, which work at a temperature of 60F (say 15C).

WEIGHT OF WATER

1 cubic foot of fresh water weighs 62.99lb = 6.299 gallons
1 gallon = 10 lb
Therefore 6.299 barrels = 1 metric tonne
36 cu.ft. = 1 ton = 224 gallons
1 cubic foot of salt water weighs 64 lb and therefore
    35 cu.ft. = 1 ton

IN THE KITCHEN

2 saltspoons = 1 teaspoon
2 teaspoons = 1 dessertspoon
2 dessertspoons = 1 tablespoon
4 tablespoons = 1 teacup
2 teacups = 1 breakfastcup
2 breakfastcups = 1 pint
1 pint   = 20 fl.oz
1 breakfastcup = 10fl.oz
1 teacup = 5 fl.oz
1 tablespoon = $1^{1}/_{4}$ fl.oz
1 dessertspoon = $^{5}/_{8}$ ths fl.oz
1 teaspoon = (say) $^{1}/_{3}$rd fl.oz
1 saltspoon = (say) $^{1}/_{6}$th fl.oz

UK measures	US measures
1 teaspoon = $^{5}/_{16}$th fl.oz	6 teaspoons = 1 liquid oz
1 tablespoon = $1^{1}/_{4}$ fl.oz	2 tablespoons = 1 liquid oz
$^{1}/_{4}$ pint = 5 fl.oz	
$^{1}/_{2}$ pint = 10 fl.oz	1 cup = 8 liquid oz
$^{3}/_{4}$ pint = 15 fl.oz	
1 pint = 20 fl.oz	1 US pint = 16 liquid oz

WINE

1 bottle = $^{1}/_{6}{}^{th}$ of 1 gallon (75.7682cl)
1 magnum = 2 bottles
1 Jeroboam = 4 bottles
1 Rehoboam = 6 bottles (1 gallon)
1 Methuselah = 8 bottles
1 Salmanazar = 12 bottles
1 Balthazar = 16 bottles
1 Nebuchadnezzar = 20 bottles
1 puncheon = 2 barrels (72 gallons)
1 butt = 3 barrels (108 gallons)
1 tun = 7 barrels (252 gallons)

BEER AND ALE

1 jug of ale = 1 pint
1 tankard = 1 quart
1 flagon or pitcher = 2 quarts = $^{1}/_{2}$ gallon
1 firkin = 9 gallons
1 kilderkin = 2 firkins
1 barrel = 4 firkins
1 hogshead = 3 kilderkins = 6 firkins = 54 gallons

APOTHECARIES' FLUID MEASURES

1 fluid scruple = 20 minims
1 drachm = 3 scruples
1 fluid ounce = 8 fluid drachms = 24 fluid scruples
1 pint = 20 fluid ounces
1 corbyn (rare) = 40 fluid ounces = 2 pints

Medicine Measures

note: these differ from In the Kitchen!

1 teaspoon = 1 fluid drachm = 60 drops or minims
1 dessertspoon = 2 fluid drachms
1 tablespoon = 4 fluid drachms
1 wineglass = 2 fluid ounces
1 teacup = 3 fluid ounces

One quart of sifted flour .... 1 pound.
One quart of corn meal .... 1 pound 2 ounces.
One pint of butter closely packed .... 1 pound.
One quart of powdered sugar .... 1 pound 7 ounces.
One quart of granulated sugar .... 1 pound 9 ounces.
A piece of butter the size of an egg ...... 2 ounces.
The white of a common sized egg weighs . 1 ounce.
Ten eggs are equal to .... 1 pound.
A common sized tumbler holds .... ½ a pint.
A common wine glass holds .... ½ a gill.

LIQUIDS.

Four tablespoonfuls .... ½ a gill.
Eight tablespoonfuls .... 1 gill.
Two gills .... ½ a pint.
Four gills .... 1 pint.
Two pints .... 1 quart.
Four quarts .... 1 gallon.
Four teacups of liquid .... 1 quart.

A page from the Columbian Recipes Methodist Cookbook of 1893, Ohio.

Paper folding into leaves and pages

Name	Abbreviation pages	Folded into leaves	Folded into
Folio	Fo	2	4
Quarto	4to	4	8
Octavo	8vo	8	16
Duodecimo	12mo	12	24
Sextodecimo	16mo	16	32
Octodecimo	18mo	18	36
Vicesimo	20mo	20	40
Vicesimoquarto	24mo	24	48
Duottrecesimo	32mo	32	64

BOOK SIZES

Name	height x width in inches	Abbreviation
Imperial Folio	22×15	Impfol
Super Royal Folio	$20 \times 13^{1}/_{2}$	suRfol
Royal Folio	$20 \times 12^{1}/_{2}$	Rfol
Medium Folio	$18 \times 11^{1}/_{2}$	Mfol
Demy Folio	$17^{1}/_{2} \times 11^{1}/_{4}$	Dfol
Crown Folio	15×10	fol
Post Folio	$15^{1}/_{4} \times 9^{1}/_{2}$	Post fol
Foolscap Folio	$13^{1}/_{2} \times 8^{1}/_{2}$	Ffol
Pott Folio	$12^{1}/_{2} \times 7^{3}/_{4}$	Pottfol
Imperial Quarto (4to)	15×11	Imp4
Super Royal 4to	$13^{1}/_{2} \times 10$	suR4
Royal 4to	$12^{1}/_{2} \times 10$	R4
Medium 4to	$11^{1}/_{2} \times 9$	M4
Demy 4to	$11^{1}/_{4} \times 8^{3}/_{4}$	D4
Crown 4to	$10 \times 7^{1}/_{2}$	C4
Post 4to	$9^{1}/_{2} \times 7^{5}/_{8}$	Post8
Foolscap 4to	$8^{1}/_{2} \times 6^{3}/_{4}$	F4
Pott 4to	$7^{3}/_{4} \times 6^{1}/_{4}$	Pott4
Imperial Octavo (8vo)	$11 \times 7^{1}/_{2}$	Imp8
Super Royal 8vo	$10 \times 6^{3}/_{4}$	suR8
Royal 8vo	$10 \times 6^{1}/_{4}$	R8
Medium	$8vo 9 \times 5^{3}/_{4}$	M8
Demy 8vo	$8^{3}/_{4} \times 5^{5}/_{8}$	D8
Large Crown 8vo	$8 \times 5^{1}/_{4}$	
Crown 8vo	$7^{1}/_{2} \times 5$	C8
Post 8vo	$7^{5}/_{8} \times 4^{7}/_{8}$	Post8
Foolscap	$6^{3}/_{4} \times 4^{1}/_{4}$	F8
Pott 8vo	$6^{1}/_{4} \times 3^{7}/_{8}$	Pott8
Demy 16mo	$5^{5}/_{8} \times 4^{3}/_{8}$	D16
Demy 18mo	$5^{3}/_{4} \times 3^{3}/_{4}$	D18
Music	$14 \times 10^{1}/_{4}$	

Books whose width exceeds their height are marked 'ob':
e.g. 'ob Crown Folio (obCfol)' is 10" high x 15" wide.

PAPER SIZES

little used now but of great antiquarian value

sizes in inches... Name	Writing & Drawing	Printing	Wrapping
Emperor	72×48		
Double Quad Crown		60×40	
Quad Imperial			58×45
Double Nicanee			56×45
Quad Royal			50×40
Antiquarian	53×31		
Casing			46×36
Saddleback			45×36
Quad Demy		45×35	
Double Imperial			45×29
Grand Eagle	$42 \times 28^{3}/_{4}$		
Quad Crown		40×30	
Double Elephant	$40 \times 26^{3}/_{4}$		
Double Royal		40×25	
Colombier	$34^{1}/_{2} \times 23^{1}/_{2}$		
Atlas	34×26		
Double Large Post	33×21	33×21	
Six Pound Grocers			32×22
Double Four Pound			31×21
Double Demy	31×20	$35 \times 22^{1}/_{2}$	
Double Post	$30^{1}/_{2} \times 19$	$31^{1}/_{2} \times 19^{1}/_{2}$	
Imperial	30×22	30×22	
Double Crown		30×20	
Imperial Cap			29×22
Elephant	28×23	28×33	34×24
Super Royal	27×19	$27^{1}/_{2} \times 20^{1}/_{2}$	
Double Foolscap	$26^{1}/_{2} \times 16^{1}/_{2}$	27×17	
Cartridge	26×21		
Haven Cap			26×21

sizes in inches... Name	Writing & Drawing	Printing	Wrapping
Four Pound Grocers			26×20
Royal	24×19	25×20	
Sheet & half Foolscap	$24^{1}/_{2} \times 13^{1}/_{2}$		
Bag Cap			$24 \times 19^{1}/_{2}$
Sheet & half Post		$23^{1}/_{2} \times 19^{1}/_{2}$	
Medium	$22 \times 17^{1}/_{2}$	23×18	
Sheet & Third Foolscap	$22 \times 13^{1}/_{2}$		
Kent Cap			21×18
Large Post	$21 \times 16^{1}/_{2}$		
Copy or Draft	20×16		$20 \times 16^{1}/_{2}$
Demy	$20 \times 15^{1}/_{2}$	$22^{1}/_{2} \times 17^{1}/_{2}$	
Music Demy		$20 \times 15^{1}/_{2}$	
Crown		20×15	
Post	$19 \times 15^{1}/_{2}$	$19^{1}/_{2} \times 15^{1}/_{2}$	
Pinched Post	$18^{1}/_{2} \times 14^{3}/_{4}$		
Foolscap	$17 \times 13^{1}/_{2}$	$17 \times 13^{1}/_{2}$	
Brief	$16^{1}/_{2} \times 13^{1}/_{2}$		
Pott	$15 \times 12^{1}/_{2}$		

Oliver Cromwell, when asked whether the water-mark of the king's head should remain on state papers, replied that "the old fool's head" could continue to be used. But 'head' was thought indelicate as it had been chopped off; hence 'foolscap'

MUSIC

1 Semibreve = 2 minims, 4 crochets, 8 quavers,
16 semiquavers, 32 demisemiquavers,
64 hemidemisemiquavers.

This sequence of numbers is compatible with the imperial system of measures but alien to the metric system. There are 12 semitones to an octave, as is shown by the fact that the notes on a piano are in groups of 7 white notes together with 5 black ones and consequently the frequency of each note is 1.0594631 (the 12th root of 2) times the frequency of the next lower one.

The ratio of minor tone to major tone is 80:81 and is called the 'comma of Pythagoras'; but, with a 'well-tempered' keyboard – as required to perform Bach's 48 preludes and fugues which were written in all the keys imaginable – the distinction between major and minor tones disappears. It was Pythagoras who discovered the basic relation between mathematics and musical harmony, from the fact that the vibration of a single stretched string produces a ground note, the notes that sound harmonious being produced by dividing the string into an exact number of parts – into exactly two parts or exactly three or four parts. If the still point on the string (the *node*) does not come at one of those exact points, the sound is discordant. As the node is shifted along the string, recognizably harmonious notes are sounded as the prescribed points are passed. From the ground note of the whole string, then with the node at the midpoint the note is one octave higher; at a point one third of the way along the note is a fifth higher; and so on.

It was Pythagoras again who found that the chords which sounded most pleasing to the Western ear correspond to exact divisions of the string by whole numbers. His followers believed that the agreement between nature and number was so significant that even the movement of celestial bodies could be calculated in relation to musical intervals. [Anne Macaulay showed – see Megalithic measures – that the geometry of the great Stone Circles of 5,000 years and more ago corresponds to this musical mathematics.]

There are no more 5s or 10s in the measurement of music than in the measurement of time. Halving and doubling is how the human brain works. Favourite popular time signatures are $^3/_4$, $^4/_4$ and $^6/_8$. The essence of the binary system of numbering, which governs computer arithmetic – the basis of modern technology – is counting in twos, not tens. [See Appendix X]

A Note on Organs *with thanks to Harry Coles*

Whether an organ is destined for Oswestry, Oslo or Osaka, its No. 1 Diapason on the Great (Principal) will speak at 8ft pitch. The imperial foot has ruled since before the monk Wulstan (d. 963) played on Winchester Cathedral's organ. If one draws a stop labelled 8ft and starts playing on its appropriate manual, its pitch – the number of vibrations per second for any given note (fast ones producing high pitch or frequency, slow ones low) – is the same as when playing the piano. Now by a quirk of natural physics, the note C (the third white note up from the bottom on a standard keyboard) is produced by an ordinary wood or metal organ pipe from its mouth to its top, just 16ft in length.

Twelve pipes on to the next C, and they have tapered down to 8ft; another twelve and that pipe is only 4ft, and so on. This is yet another illustration of the natural progression or regression of numbers, which is wholly at odds with decimal arithmetic.

Some pipes for very high notes are only a few inches long. A large organ such as in Liverpool's modern Anglican Cathedral contains thousands of pipes, all fashioned to precise specifications. In contrast, a Victorian 'Father Willis' church organ has stops of 16, 8, 4 and 2 ft pitches – and so will any similar organ, whether in Tokyo, Timbuktu or Tewkesbury!

SOUND

The speed of sound in air varies with temperature. It's about 1,125ft per second at 68°F but just under 1,087 ft/sec at 32°F. The normal standard is 1,100 ft/sec which is exactly 1 mile in 4.8 seconds. Sound travels much faster in water, at about 4,760ft per second, or 1 mile in about 1.1 seconds.

STRESS AND PRESSURE

Different scales are used to measure pressure. A barometer balances atmospheric pressure against the pressure exerted by a column of mercury whose height is measured in inches. Normal atmospheric pressure at sea level is about 30 inches of mercury, which is adopted as the British standard. The standard in the metric system is defined as 101.325 kilopascals which comes to very nearly 760mm of mercury.

But 30in = 762mm; whereas 101.325 kilopascals corresponds to a pressure of very nearly 2,116.22lb/sq.ft or 14.6959lb/sq.in. *Tyre* pressures are measured in lb per sq.in.; but saying that the air pressure in a tyre is 25lb per sq.in. means that this is the amount by which that pressure *exceeds* atmospheric pressure. *Blood* pressure is measured in the UK in mm of mercury (whereas the French use cm – which certainly raises the blood pressure!).

Now, how are these three scales related? Well, 1lb/sq.in. is *very* nearly 51.7149 millimetres or 2.03602 inches of mercury.

Weather forecasters usually give atmospheric pressure in millibars (1 millibar = 100 pascals), in spite of the fact that the unit used on many thousands of barometers up and down the country is the inch of mercury. However, the stress in a wire rope may well be measured in tons per square inch. The pound per square inch is probably the most familiar of these units, so here are the values of the others in terms of that one.

1 kilopascal	=	0.145038 pounds per square inch
1 millibar	=	0.0145038 pounds per square inch
1mm of mercury	=	0.0193368 pounds per square inch
1in of mercury	=	0.491154 pounds per square inch
1 ton per sq in	=	2,240 pounds per square inch

TEMPERATURE

The Fahrenheit scale, named after Gabriel Daniel Fahrenheit (1686-1736), a German physicist who, at first, took 100° to represent the temperature of his own body, which he later discovered to be about 98.4°. There are two theories as to how he chose 0 on his scale: either it was on an extremely cold day in Gdansk in 1709 or it was the temperature of a mixture of salt and ice.

The modern keys to the Fahrenheit scale are: absolute zero, which is *minus* 459.67°, and the temperature of the *triple point* of water – at which ice, water and water vapour can co-exist – which is 32.018°F.

The temperature of the triple point of water on the Kelvin scale is 273.16°; this number being chosen so that the size of the kelvin should match the Celsius degree. The Rankine scale does for Fahrenheit what the Kelvin scale does for Celsius. The Kelvin scale, named after William Thomson, Lord Kelvin (1824-1907), is based on thermodynamic principles. Fahrenheit, Rankine and Celsius are each given in degrees, but kelvin is just kelvin.

The following table shows a sample of temperatures, on these four scales, relating to: absolute zero, freezing point of water, normal body temperature, and boiling point of water at standard pressure.

kelvin	°Celsius	°Fahrenheit	°Rankine
0	- 273.15	- 459.67	0
273.15	0	32	491.67
310.05	36.9	98.42	558.09
373.15	100	212	671.67

Incidentally, it is a vital principle of the metric system that only one unit is authorized for each type of magnitude. Therefore, since the kelvin is the metric unit of temperature, the degree Celsius does not belong to the metric system.

FORCE

While the concept of *force* as 'mass × acceleration' was understood long before the advent of the metric system, its importance was not appreciated either side of the Channel. So the same name (pound or kilogram) was used for both mass and weight. [See also Weight and Mass] But once machinery with fast-moving parts appeared, the distinction became very important.

The *weight* of components had to be divided by 'g' in order to calculate the necessary forces of acceleration. And with the new knowledge of electricity, it became extremely important, because this used forces measurable by neither mass nor weight – such as the powerful force between two wires of negligible weight carrying a current of zero mass!

The *poundal* is the force which, applied to a particle of mass 1 pound, gives it an acceleration of 1 ft. per second per second. If a particle starts from rest, accelerating at that rate, it is moving at a speed of 1ft per second after 1 second, 2ft per second after 2 seconds, 3ft per second after 3 seconds, and so on. A particle of mass 1lb falling freely (i.e. acted upon only by its own weight) has an acceleration of about 32.187ft/sec/sec; so a particle of mass 1 pound has a weight of about 32.187 *poundals*. (The unit of force in the metric system is defined in the same way. The newton is the force which, applied to a particle of mass 1 kilogram, gives it an acceleration of 1 metre/sec/sec.) The aero industry likewise devised a new unit of mass called the *slug* = 32.187lb.

The force equal to the weight of a particle whose mass is 1lb is called the lb weight or pound force. So 1 poundal = 1/32.187lb *force*. In practice, the word weight or force is usually omitted, so confusion often arises. The British system thus has two basic units of force: the poundal and the pound weight. One is based on *inertia* and the other on *gravity*. While the lb weight is used for everyday purposes, the poundal is preferred for scientific work – partly because weight varies slightly with location on the planet.

The approximate weight of a particle of mass 1lb at different latitudes is shown in the following table, which demonstrates the relevance of references to latitude in the section on work and energy.

latitude (degrees)	weight (poundals)
0	32.0877
15	32.099
30	32.1301
45	32.1726
60	32.2152
75	32.2464
90	32.2578

The standard (legal) formula for converting pounds weight to poundals is: 6,096lb weight = 196,133 poundals. This gives the weight of a particle of mass 1lb at a latitude of about $45\frac{1}{2}°$. [see Appendix XVII]

WORK AND ENERGY

To raise a particle of mass 1lb by 1ft, 1ft.lb weight (or 1 foot pound) of energy is needed. Likewise, if something is moved 1 ft against a resisting force of 1 poundal, then 1 foot poundal of work is done.

Following the discovery that heat is a form of energy (the 1st law of thermodynamics), the English physicist James Prescott Joule* (1818-1889) – working, of course, in English units – found that, in the latitude of Manchester, 1 British Thermal Unit equalled 773.64ft.lb. The legal conversion formula is: 7.4726673 BTU = 5,815ft.lb.

1 BTU represents the amount of heat required to raise the temperature of 1 lb of water at its greatest density (occurring at about 39°F) by 1°F. Thus, the heat required to raise the temperature of 1 pint of water ($1^{1}/_{4}$ lb) from 52°F to boiling point – a rise of 160°– is $160 \times 1^{1}/_{4} = 200$ BTU. The *therm* = 100,000 BTU.

*The *joule* is the *Systeme International* unit of work and energy, replacing the calorie – which is *not* a metric unit, even though nutritionists use it with impunity. Difficulties arise because the calorie is a unit of heat whereas the joule is a unit of mechanical energy. The joule is defined as the work done (energy transferred) by a force of one newton acting over one metre. The calorie is defined as 4.1868 joules and the BTU as very nearly 1.05506 kilojoules. [see Appendix XVII]

POWER

Power is the rate at which energy is converted from one form into another. In the metric system, the unit of power is the *watt*, named after the Scottish engineer James Watt (1736-1819). 1 watt is a rate of working of 1 joule per second: e.g., a 1 kilowatt electric fire turns electrical energy into heat at the rate of 1,000 joules per second. Before the advent of the steam engine, motive power was usually supplied by horses. So, when James Watt sold his steam engines, he had to tell his customers how many horses they would replace. He found that a horse would walk comfortably at $2^1/_2$ m.p.h. while pulling against a resistance of 100lb. Now, $2^1/_2$ m.p.h. = 13,200 ft. in 3,600 seconds = 3.660ft/sec. So the horse was working at a rate of 366.66ft.lb/sec. Watt didn't want to overstate his case, so he added a half again to this figure, giving 550ft.lb/sec. That, then, is the British horse-power.

The Continentals decided on 75 metre kilograms/sec. for their *cheval-vapeur*. So the British HP = 745.69987158227022 watts (the official government figure and the longest in this book), whereas the European HP = a mere 735.49875 watts.

A machine HP is usually reckoned to equal the power of 4.4 horses, one horse being assumed to possess the strength of 5 men; so 1 HP notionally matches the strength of 22 men. But that is entirely fanciful. For if an 11 stone man pedals a 6lb bicycle up a 1 in 20 hill at 10mph he is doing work at the rate of more than $((11 \times 14) + 6) \times 10 \times {}^{22}/_{15} \times 20$ft.lbs/sec = 117.33ft.lbs/sec or 0.2133 HP – which is nearly 5 times the rate of working suggested by 22 men being equivalent to 1 HP!

Horse-power (as in motor-cars) is determined *either* as *indicated* HP, which is the power developed in the engine cylinders as calculated from:

a) the average pressure of the working fluid

b) the piston area

c) the stroke and

d) the number of working strokes /minute, *or* as *brake* HP, which is the actual power of the engine calculated from:

i) the force exerted on a friction brake or absorption dynamometer acting on a fly-wheel or brake-wheel rim

ii) the effective radius of this force and

iii) the speed of the fly-wheel or brake-wheel.

Imperial and metric units are often used together. Thus, it takes about 1.023 megajoules to convert 1lb of water to 1lb of steam at 212°F. To take an example at random, 35.29 megajoules are needed for 34½ lb, and if this is done in 3,600 seconds, the rate of working is 9.802 kilowatts or 13.15 horse-power. This mixing is occasionally convenient but often dangerous.

SPEED

One of the many beauties of the imperial system is the compatibility between its units of distance and the measurement of time, because of the duodecimal factor common to both: e.g.
1 m.p.h. = 1,056in/minute = 88ft/minute;
and 1 mile per day (24 hours) = 220ft/hr., etc.

A Mach number, named after Ernst Mach (1838-1916) the Austrian physicist, indicates the speed ratio of an object to the speed of sound – about 1,087 feet per second through air at sea level at 32°F. So Mach 1 equals the speed of sound; while hypersonic speed means Mach 5 or faster.

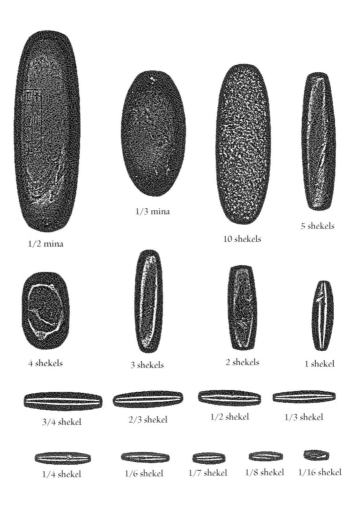

1/2 mina

1/3 mina

10 shekels

5 shekels

4 shekels

3 shekels

2 shekels

1 shekel

3/4 shekel

2/3 shekel

1/2 shekel

1/3 shekel

1/4 shekel

1/6 shekel

1/7 shekel

1/8 shekel

1/16 shekel

A complete set of weights for measuring precious metals in the palace (see page 45). Babylon, c. 1800 BC. Drawn from the collection in the British Museum.

An Anthology

Then down with every metric scheme
Taught by the foreign school;
We'll worship still our father's God
And keep our father's rule –
A perfect inch, a perfect pint,
The honest British pound,
Shall hold their place upon the Earth
Till time's last trump shall sound

from a popular 19th century song

Economists have come to feel
What can't be measured isn't real
The truth is always an amount
Count numbers, only numbers count.

Robert Chambers

An old-timer's explanation for the 1980s Australian rural recession (anon)

It all started in the sixties when they changed pounds to dollars: that doubled me overdraft. Then they brought in kilograms instead of pounds: me wool-clip dropped by half. After that they changed rain to millimetres: and we haven't had an inch of rain since. If that wasn't enough, they brought in Celsius, so we never get above forty degrees – no wonder the wheat don't grow. Next came hectares instead of acres; so I only had half the land I used to own. By this time I'd had enough, and decided to sell out. I put the property into the agent's hands, but then they changed miles into kilometres, so now I'm twice as far out of town and nobody wants to buy the b----- place!

An ounce is to dangle from your finger
A pound is to hold in your hand
A stone is to carry under your arm
A hundredweight over your shoulder
And a ton a horse pulls in a cart

traditional

The woods are lovely, dark and deep.
But I have promises to keep.
And miles to go before I sleep,
And miles to go before I sleep.

Robert Frost, 1874-1963

FROM THE BIBLE

Leviticus XIX 36: Just balances, just weights, a just epha, and a just hin, shall ye have: [*the 'epha' or 'ephah' was a Hebrew dry measure, corresponding to the 'bath' – see below – while the 'hin' was a liquid measure – containing a little over a gallon – see below*]

Isaiah V 10: Yea, ten acres of vineyard shall yield one bath, and the seed of an homer shall yield an ephah. [*1 bath measured nearly 8 gallons (1 bushel) of wine; and 1 homer equalled 10 ephahs (say 9 bushels). So, unless the Israelites mended their ways, 1 acre would produce less than 1 gallon of wine – barely $1/10^{th}$ of normal, while 10 measures of seed would produce only 1 measure of crop. $1/6^{th}$ of a bath (nearly 11 pints) was called a 'hin' which was divided into 12 'logs', so 1 log equalled nearly 1 pint.*]

Samuel I XIV 14: And that first slaughter, which Jonathan and his armour-bearer made, was about twenty men, within an half-acre of land, which a yoke of oxen might plough.

Kings I X 17: And he made three hundred shields of beaten gold; three pound of gold went to one shield.

Ezra II 69: They gave after their ability unto the treasure of the work three-score and one thousand drams of gold, and five thousand pounds of silver, and one hundred priests' garments.

Nehemiah VII 71-72: And some of the chief of the fathers gave to the treasure of the work twenty thousand drams of gold, and two thousand and two hundred pounds of silver. And that which the rest of the people gave was twenty thousand drams of gold, and two thousand pounds of silver, and three-score and seven priests' garments. [*'pound' is translated from the original 'maneh', which is believed to have equalled about 1¹/₄ lb; and 60 manehs equalled 1 'talent', i.e. say 75lb*]

Isaiah XL 12 and 21-22: Who hath measured the waters in the hollow of his hand, and meted out heaven with the span, and comprehended the dust of the earth in a measure, and weighed the mountains in scales and the hills in a balance?

Daniel V 25: Mene, Mene, Tekel, Upharsin. This is the interpretation of the thing: Mene – God hath numbered thy kingdom, and finished it; Tekel – Thou art weighed in the balances and art found wanting.

Zechariah II 1-2: I lifted up my eyes again, and looked, and behold a man with a measuring line in his hand. Then said I, Whither goest thou? And he said unto me, To measure Jerusalem, to see what is the breadth thereof, and what is the length thereof.

Job XXXIIX 5-7: Who hath laid the measures thereof, if thou knowest? Or who hath stretched the line upon it? Whereupon are the foundations thereof fastened?

Ezekiel XL 5 and XLIII 13-15: And behold a wall on the outside of the house round about, and in the man's hand a measuring reed of six cubits long by the cubit and an hand breadth: so he measured the breadth of the building, one reed; and the height, one reed. And these are the measures of the altar after the cubits: The cubit is a cubit and a hand breadth; even the bottom shall be a cubit, and the breadth a cubit, and the border thereof by the edge thereof round about shall be a span: and this shall be the higher place of the altar. And from the bottom upon the ground even to the lower settle shall be two cubits, and the breadth one cubit; and from the lesser settle even to the greater settle shall be four cubits, and the breadth one cubit. So the altar shall be four cubits; and from the altar and upward shall be four horns. [*Yes, most obscure!*]

Proverbs xx 10: Divers weights and divers measures...

John ii 6: And there were set there six water-pots of stone, after the manner of the purifying of the Jews, containing two or three firkins apiece.

John vi 19: So when they had rowed about five and twenty or thirty furlongs, they see Jesus walking on the sea, and drawing nigh unto the ship; and they were afraid.

John xi 18: Now Bethany was nigh unto Jerusalem, about fifteen furlongs off.

Luke xxiv 13: And, behold, two of them went that same day to a village called Emmaus, which was from Jerusalem about three-score furlongs. [*i.e. 7^1/$_2$ miles*]

Revelation xxi 16: And the city lieth foursquare, and the length is as large as the breadth; and he measured the city with the reed, twelve thousand furlongs. The length and the breadth and the height of it are equal.

Matthew v 41: Whosoever shall compel thee to go a mile, go with him twain.

Many thanks to Tony Bennett for his Biblical research

OTHERS (CHRONOLOGICALLY)

Hesiod (9th century BC): Hew a mortar three feet in diameter, and a pestle three cubits, and an axle-tree seven feet...and hew a wheel of three spans for the plough-carriage of ten palms.

Herodotus (BC484-424): All men who are short of land measure it by fathoms; but those who are less short of it, by stadia; and those who have much, by parasangs; and such as have a very great extent, by schoinoi. Now a parasang is equal to 30 stadia, and each schoinos, which is an Egyptian measure, is equal to 60 stadia.

Plato (BC c428-347): The most important and first study is of numbers and the greatest of their influence on the nature of reality...Every diagram and system of number and every combination of harmony and the agreement of the revolution of the stars must be made manifest as one in all to him who learns in the proper way, and will be made manifest if a person learns aright by keeping his eyes on unity...There is a natural bond linking all things.

Cicero (BC106-43). Not to know what happened before we were born is to remain perpetually a child. For what is the worth of a human life unless it is woven into the life of our ancestors by the records of history?

Flavius Josephus (AD c37-100): The author of weights and measures, an innovation that changed a world of innocent and noble simplicity, in which people had hitherto lived without such systems, into one forever filled with dishonesty.

Leon Battista Alberti (1440): We shall ever give ground to honour. It will stand to us like a public accountant; just, practical, and prudent in measuring, weighing, considering, evaluating, and assessing everything we do....

St Isadore of Seville (c1600): Take away number in all things and all things perish. Take calculation from the world and all is enveloped in dark ignorance; nor can he who does not know the way to reckon be distinguished from the rest of the animals.

Milton (1608-74): In his hand / He took the golden Compasses, prepar'd / In God's eternal store, to circumscribe / This universe, and all created things: / One foot he center'd, and the other turn'd / Round through the vast profunditie obscure / And said, thus far extend / Thus far thy bounds / This be thy just Circumference, O World. (Paradise Lost, Book VI)

Napoleon Buonaparte (1769-1821): The scientists adopted the decimal system on the basis of the metre as a unit. Nothing is more contrary to the organisation of the mind, memory and imagination. The new system will be a stumbling block and source of difficulties for generations to come. It is just tormenting the people with trivia.

Robert Southey (1774-1843): An ounce of love is worth a pound of knowledge. (Life of Wesley: ch.16)

Charlotte Elliott (1789-1871): Just as I am, of that free love, The breadth, length, depth and height to prove,

Lord Kelvin (1824 -1907): I often say, when you can measure what you are speaking about and express it in numbers, you know something about it; but when you cannot measure it, when you cannot express it in numbers, your knowledge is of a meagre and unsatisfactory kind.

Lewis Carroll (1832-98): Rule forty-two. All persons more than a mile high to leave the Court. (Alice in Wonderland: ch.12) [see also Appendix VIII]

W H Auden (1907-73): And still they come, new from those nations to whom the study of that which can be weighed and measured is a consuming love.

Eric A Blair / George Orwell (1903-50): "And what in hell's name is a pint?" said the barman, leaning forward with the tips of his fingers on the counter. "Ark at 'im! Calls 'isself a barman and don't know what a pint is! Why, a pint's the 'alf of a quart, and there's four quarts to the gallon. 'Ave to teach you the ABC next." "Never heard of 'em", said the barman shortly. "Litre and half litre – that's all we serve." *(Nineteen Eighty-Four* 1949)

MEGALITHIC MEASURES

The origins of imperial measures derive from mankind's earliest knowledge of astronomy and geometry, of music and the measurement of time. But before the discovery of any means of recording such knowledge and transmitting it to succeeding generations, the rudiments of these sciences cannot be traced farther back than the late Neolithic period of around 4,000BC. This was otherwise known as the Megalithic period, because it was distinguished chiefly by the creation of the world's earliest structures, in the shape of the great stone rings, which were erected throughout what are now the British Isles and NW France.

The remains of some 900 are still standing, of which some 200 of the most prominent were meticulously surveyed by Alexander Thom (Professor of Engineering Science at Oxford University from 1945 to 1961) and his son A S Thom (senior lecturer in the Department of Aeronautics and Fluid Mechanics at Glasgow University). Their analyses were published in three major books (OUP) from 1967 to 1978. They proved the astronomical purposes for which several of these colossal monuments were built and also that they were all planned and engineered on the basis of two key units of length: the *Megalithic Yard* equal to 2.72ft and the *Megalithic Rod* of 6.8ft; so that 5 MY = 2 MR. They identified, furthermore, many distinct types of geometric layout.

Thoms' work has been vastly extended and elaborated by Anne Macaulay (1924-1998), the results of whose research, throughout the last thirty years of her life, were published in a book in 2006. She showed that these Megalithic people on the Atlantic seaboard virtually invented mathematics: whether even more ancient civilizations in China had done so independently is another question. For these ancient Britons must have devised the Fibonacci series of numbers (commonly called a series although more properly a sequence) almost 5,000 years before the birth of Leonardo of Pisa (c.1170-1250), known as Fibonacci – the man who introduced Arabic numerals to Europe.

This series is an additive progression, beginning with 0, 1, 1, 2, 3, 5, 8, 13, 21, 34, 55, etc, where each term after the first two is the sum of the two immediately preceding terms. It has numerous fascinating and beautiful properties, and not only governs the ratios of many dimensions in Megalithic geometry, but also those in many natural phenomena: e.g. the laws governing the multi-reflection of light by mirrors, the rhythmic laws of gains and losses in the radiation of energy, the ratio of males to females in honey bee hives and the breeding pattern of rabbits – a symbol of fecundity.

They must also have discovered the use of square roots, deriving from ratios used in planning many of the layouts. Thus, the ratios of the base of an equilateral triangle to its height and of the side to the diagonal in a 1:2 rectangle (1/square roots of 3 and 5), etc. In the Pythagorean number cosmology of ancient Greece, these incommensurable ratios symbolize the immeasurable archetypes that are *a priori* to the visible world of form. The Pythagorean school visualized an organizational foundation for the universe, emerging from invisible vibrational patterns and numerical proportions into the visible world of shape and form. So Megalithic proto-Britons and Bretons were employing Pythagorean mathematics 2,000 years before the birth of Pythagoras.

Yet again, as Anne Macaulay proved, they must have discovered the Golden Mean, the most intriguing and useful of these incommensurable proportional relationships, springing from the pentagon. It is represented by 1.618034, symbolized by *phi* – and most easily recognized as the ratio of each of the longer sides to the shortest side in a triangle in which the angle at the narrow end is 36° and each of the other two angles is 72°. The smaller part relates to the larger precisely as the larger relates to the whole. Likewise, each of the triangles in a pentagonal star has two equal sides that relate to the third side as 8 to 5, the most rudimentary basis of the Golden Mean. *Phi* governs universal laws of proportion: discernible in the fuselage/wingspan ratio of a jumbo-jet, the features of classical architecture and the proportions of the human frame – as famously illustrated in Da Vinci's drawing of Vitruvian Man – and, most naturally and precisely, in the spiralled forms of marine shells and the heads of certain plants, particularly sunflowers, by an arrangement called phyllotaxis.

A typical sunflower head may have 34 and 55 opposing spirals – producing a ratio of 0.6181818 – or 89 and 144 (0.6180555) or 144 and 233 (0.6180257), etc. As György Doczi said (Ref: Shambhala Publications Inc, Boulder, Colorado 80302, 1981):

One does not to have to fear vengeful deities to feel something like awe at such unexpected precision in a pattern of natural growth. It seems unreasonable to believe that the number of seeds in a sunflower is pre ordained, yet something like that is exactly what happens. Irrational numbers are not unreasonable; they are only beyond reason, in the sense that they are beyond the grasp of whole numbers. Like pi – 3.1415926 etc – phi can be calculated to endless series of decimal places without ever reaching a rational number. They are infinite and intangible. In patterns of organic growth the irrational phi ratio of the golden section reveals that there is indeed an irrational and intangible side to our world.

As William Blake wrote:

> *To see a world in a grain of sand*
> *and Heaven in a wild flower*
> *hold infinity in the palm of your hand*
> *and eternity in an hour.*

Phi is normally specified as 'y'/2 where 'y' = the square root of 5 plus 1. Thus, (2.236068 + 1) / 2 = 1.618034. However, *phi* is often invoked by its reciprocal: (2.236068 – 1) / 2 = 0.618034.

Note that: $1.618034 \times 0.618034 = 1.00$
whilst $1.618034 \times 1.618034 = 2.618034$

In a Fibonacci series, therefore, each new number – the sum of the previous two – divided by the last one, produces a number close to *phi*. Also, the greater the numbers, the closer the quotient grows towards *phi*.

So, $^5/_3 = 1.66667$, $^{21}/_{13} = 1.615385$, $^{144}/_{89} = 1.617978$,
$^{233}/_{144} = 1.618056\ldots$
(The same symmetrical and harmonic patterns have been recognized in modern times in atomic structures, quantum and wave mechanics, hydrodynamics, electrochemical reactions and molecular bonding.) Anne Macaulay realized – by one of her many inspirations – the direct relationship between this divine ratio and our system of customary measures:

6 MR (40.8 ft.) $\times$ *phi* (1.618034) = 66 ft. = 4 imperial rods
= 22 yards.

This chain, therefore, stretches back some 6,000 years!

So she established that our Megalithic ancestors were competent in the Pythagorean *quadrivium* of arithmetic (number), geometry (number in space), music (number in time) and astronomy / astrology (number in space and time). Imperial measures grew from the roots of Europe's earliest civilization. To quote from Boyer and Merzbach's *A History of Mathematics* (John Wiley and Sons, 1991): 'That the beginnings of mathematics are older than the oldest civilization is clear.' And from B L Van der Waerden's classic Geometry and algebra in ancient civilizations: 'We have seen so many similarities between the mathematical and religious ideas current in neolithic England, in Greece, India and China (in the Han period), that we are bound to postulate the existence of a common metrological doctrine from which all these ideas derived.'

It is this doctrine, the very heart of human culture, that those intent on imposing a metric monopoly are determined to destroy.

COSMIC NUMEROLOGY AND CUSTOMARY MEASURE

from Martin Doutre's paper

ANCIENT WEIGHTS, VOLUMES AND MEASUREMENTS

published in New Zealand in 2002

"It has long been realized that there were apparent relationships and common pedigrees within systems of metrology adopted by several early and late civilizations in the Middle East, Near East, Mediterranean and Europe. Particular numbers and ratios tend to recur in grain weights, cubic capacities and length increments between cousinly civilizations. The foundation numbers for each system are complex and few researchers have any inkling as to why such cumbersome values were chosen in the first instance. Each nation encoded into their standard measures highly scientific information about such things as the equatorial circumference of the Earth, principles of navigation, and cycles of the sun and moon to maintain accurate calendar systems. The number families were primarily founded upon '6 and 7' increases, working in unison or separately, an '11' series of navigational numbers, individual cubits that related to 3 closely associated readings of the size of the Earth, and a *phi* series of numbers for special purposes.

Therefore, to understand the numbers, one must first delve into the astronomical and navigational sciences developed in antiquity and come to realize that the values incorporated into a length, weight or volume were for mnemonic reference to such things as the 'size of the Earth' or 'cycle of the moon', etc. Then the ancient standards of the Sumerians/Babylonians, Egyptians, Hebrews/Phoenicians, Greeks, Romans, Celts/Norse/Gauls/ Iberians/ Britons, and others of these peoples who scattered abroad, can be understood, restored to full integrity and shown to be of common origin. The original system covered all contingencies and used the full range of base numbers from 1 to 12, including the ratios *pi* and *phi*. For centuries archaeologists have attempted to reconstruct the regional standards of past civilizations by taking measurements within ruined remnant structures, measuring capacities within surviving amphora vessels and by analyzing the decipherable meanings within ancient inscriptions and texts. From these endeavours a reasonably healthy dossier of information now exists and we have a fair knowledge of the measurement increments, weights and volumes preferred by particular civilizations. Over time, small amounts of drift and error may well have occurred in standards between different nations, but the important factor was that their respective systems were fashioned on the self-same parcel of original base numbers.

In most instances, the final (official) figure, as to what constituted the national standard in use during a particular epoch, is based on averages.

Now, multiplying the length of the average nautical mile of 6,076.84 ft by the number of minutes in a circle, 21,600, produces a value for the length of the meridian, the great circle through the poles, of 131,259,744ft – the officially recognized length being 131,260,000ft . So there are variations, because 'averages' are dangerous and the Earth is an irregular shape."

John Michell, however, in his great book, *The Dimensions of Paradise* (Thames and Hudson, 1988) seeks to demonstrate that all cosmological distances are essentially duodecimal multiples which are common to all ancient sciences.

He takes 6,082.56 ft, the correct figure for the nautical mile, times 21,600 = 131,383,296 ft, which equals:
$12 \times 12 \times 12 \times 12 \times 12 \times 12 \times 44$ ft, or (see over...)

135,000,000	Roman feet of 0.9732096 ft
90,000,000	Roman cubits of 1.4598144 ft
216,000	Roman furlongs of 608.256 ft
27,000	Roman miles of 4,866.048 ft
129,600,000	Greek ft of 1.01376 ft
86,400,000	Greek cubits of 1.52064 ft
207,360	Greek furlongs of 633.6 ft
25,920	Greek miles of 5,068.8 ft
114,048,000	Egyptian ft of 1.152 ft
76,032,000	Egyptian cubits of 1.728 ft

Moreover, 131,383,296 ft = 24,883.2 miles which equals 12 to the power 5 divided by 10 and is decidedly closer to modern estimates from satellite data of the Earth's meridian than the estimate made of that distance by the French scientists of the 18th century for the purpose of defining the metre!

The metrologist A R Berriman wrote that, had the French accepted Cassini's proposal in 1720 for a scientific foot based on $\frac{1}{6,000}$th part of a minute of average latitude, that unit would have been recognized as identical with the longer Greek foot by which the Parthenon was built. [Though, of course, the Parthenon wasn't built at *average* latitude; and it could be argued diabolically that the 'English' mile gained some 280ft by Elizabeth I's Act of 1593 – but see Appendix VII]

However, the fundamental point made by Berriman was that 'The names, Greek, Roman and Egyptian are applied to these units by convention only, for they all belong to the same number system and represent fractions of the Earth's dimensions.'

Nevertheless, it must be conceded that these findings are open to criticisms that scant scientific evidence exists for the actual use of some of these units within their respective ancient civilizations, that it is impossible to measure any of these units to one millionth part of a foot and that this whole area of study lends itself to historical fantasy or even obsession. But, as explained in one of the articles under Standard linear measures, this is not an exercise in physical investigation but the working out of a cosmic scheme.

Besides, (a) if calculations are to be illustrated, they have to be expressed precisely, even to an extent that in practice is unrealistic, in order to demonstrate the perfect accuracy of the theory, and (b) ignorant criticism of isolated facets of this vast subject proves little, and (c) yes, the numerology of metrology does become obsessive, but so does dedication to any art or religion, and what subject is more important than the very source of civilization?

The point is that all the ever-increasing evidence that does exist supports John Michell's general conclusions about the inter-relationships of ancient units of measurement, their common derivation from a duodecimal system unified by the English foot, and their consistency with the cosmic dimensions of distance and time.

To quote Michell further: As calculated above, the longer Roman foot was equal to 0.9732096ft and the longer Greek foot to 1.01376ft. Five thousand of these units made up their respective miles, so the Roman mile was of 4,866.048ft and the

Assyrian	63		15		9	7		6															
Iberian	64								20		14		8		5	4							
Roman		16								21						49	24		14	7			
Common Egyptian												15				50						48	20
Greek/English				10														25			49		
Common Greek							8						9						15				21
Persian							7																
Belgic																							
Sumerian														6						8			
English archaic																							
Royal Egyptian															5								
Russian																							

Table by John Michell and John Neal showing fractional relationships between the most common ancient feet.

Greek mile of 5,068.8ft. The ratio between them is 24:25, and this ratio is also found to obtain between the Greek mile and the English mile of 5,280ft. The three mile units thus form a geometric progression.

Roman mile $\times\ ^{25}/_{24}$ = Greek mile
Greek mile $\times\ ^{25}/_{24}$ = English mile

It is so unlikely that this neat progression could have arisen by chance that the author feels justified in claiming confirmation of those unit values previously arrived at by other considerations. Linking the English units with those of classical metrology also justifies making the English foot the prime unit of reference.

John Neal's studies into the ancient metrologies of many nations also concluded that it was all a single system. "Without looking at a single value one could reach this conclusion from their identical anthropometric origins whereby all of the modules are related to the human body. Fingers, thumbs, palms, hands, spans, feet, cubits and fathoms and so forth. Then from the human body in motion, steps and paces, come the itinerary modules of stadia, miles and leagues."

Assyrian																
Iberian																
Roman																
Common Egyptian																
Greek/English	14		9	7	6											
Common Greek					48		24	15	9								
Persian						49				49	9						
Belgic	15							25		50		27	15				
Sumerian								16				28		24			
English archaic		10											16		35	20	
Royal Egyptian			8						10					25	36		48
Russian				7							10					21	49

Neal continues: "Because of the acknowledged variations in all of the national standards (which had mistakenly been attributed to slackness in their regard to measure) I had found it remarkably difficult to ascertain any module at all with certainty. Largely because most archaeologists were prone to take an average for their definition of the modules, thereby destroying the mathematical purpose of the variations, (which is the maintenance of integers in their designs).

Having isolated the diverse foot modules of different nations I was able to observe that they all relate by a single fractional adjustment, just four are given here:

Assyrian	15	9	5		
Roman	16			24	8
Greek		10		25	
Belgic			6		9

Roman 24 to 25 Greek, Roman 8 to 9 Belgic and so forth. These were long acknowledged relationships, but it was possible to extend this bond through all of the modules of antiquity.

My friend and colleague, John Michell, also noticed the prevalence of square numbers in the results, visible above as 9, 16, 25 and produced the tables shown (*opposite and on previous page*) as perhaps the most historically important charts yet devised toward an understanding of ancient metrology. The abandonment of this sophisticated system in favour of decimalisation is the greatest act of cultural vandalism ever perpetrated."

Modern researchers who accept the fiction that the English units are of recent origin, and adopt the habit of expressing the values of the old measures in terms of the new-fangled metre, thereby disguise from themselves the most significant aspect of ancient metrology, its basis in canonical or classic number – i.e. number that embodies knowledge.

Were it not for that aspect, the business of establishing the exact

1/2	.514285	3	½ common Greek foot
2/3	.685714	4 8	10 digit palm length of Sumerian foot
3/4	.771428	9 15	½ common Greek cubit (12 digit span, 9inch dodrans)
4/5	.822857	16 24	½ Sumerian cubit "
5/6	.857142	25 35	½ royal Egyptian cubit "
6/7	.881632	36 48	not identified (6:7 of common Greek)
7/8	.9	49 63	Assyrian foot
8/9	.914285	64 80	Iberian foot
9/10	.925714	81 99	not identified (6:7 of Belgic)
10/11	.935064	100 120	" (6:7 of Sumerian)
11/12	.942857	99 121	Samian foot
	.953281	80 100	lesser Roman foot
	.964285	63 81	not identified (6:7 of Nippur foot)
	.979591	48 64	common Egyptian foot
35	1	35 49	English – Greek "Olympic" foot
36	1.028571	36 48	common Greek foot
	1.05	49 63	Persian foot
	1.066666	64 80	not identified (7:6 of Iberian foot)
	1.08	81 99	greater Belgic foot
12/11	1.090909	100 120	Sumerian foot
11/10	1.1	99 121	Saxon foot
10/9	1.111111	80 100	archaic English foot
9/8	1.125	63 81	the foot of Nippur
8/7	1.142857	48 64	royal Egyptian foot
7/6	1.166666	35 49	Russian ½ arshin
6/5	1.2	24 36	Roman remen or 20 digit "palimpes".
5/4	1.25	15 25	Greek/English 20 digit "pygon"
4/3	1.333333	8 16	16 inch module widely used by English speaking people
3/2	1.5	3 9	English-Greek 1½ foot cubit
2/1	2	4	English-Greek 2 foot cubit

Suggestive scheme by John Michell and John Neal showing the harmonic structure of the relationships between various ancient feet. The entire scheme is based on the English foot.

values of the old units would be of merely academic interest. The fact that those units, as here calculated in English feet, exhibit the same scale of numbers as found in ancient music and geometry is what makes the system of ancient metrology so relevant to the study of traditional science.

As seen above, the English units relate numerically to the Earth's dimensions through the powers of 12: but the most direct geodetic reference to the foot is the equator. In ancient China and Babylon the circle of the equator was divided into $365\frac{1}{4}$ degrees to represent the number of days in a solar (tropical) year – now taken as 365.2422 days. Each of the $365\frac{1}{4}$

degrees would measure some 360,000ft, each minute 6,000ft and each second 100ft.

If the number of the days in the year is taken as 365.24322, these figures become exact and accord with the estimate of the equator, which is just under 131,480,000ft. Conversely, dividing the equator into 360 degrees makes each degree equal to 365,243.22ft, or the number of days in a thousand years.

Another relic of the archaic tradition that produced these divisions of time is our present system of measurement by units of feet, furlongs and miles, with the acre as the unit of land measuring. Those measures, which are still found the most convenient today, were canonized and held sacred, because not only do they relate both to the human and to the astronomical scale, expressing the unity between the macrocosm and the microcosm, but they bring out the same numbers in the dimensions of the solar system as were given to the units of time. canonical dimensions of the earth, sun and moon, are:

Diameter of the sun 864,000 miles	$(12 \times 12 \times 6,000)$
Radius of the sun 432,000 miles	$(12 \times 12 \times 3,000)$
Diameter of the moon 2,160 miles	$(6 \times 6 \times 60)$
Radius of the moon 1,080 miles	$(6 \times 6 \times 30)$
Mean diameter of earth 7,920 miles	$(12 \times 12 \times 55)$
Mean circumference of the earth	24,883.2 miles
	(12x12x12x12x1.2)
Circumference of the moon	12^7 feet
Distance from earth to moon	237,600 miles
(or 60 × the Earth's radius)	$(6 \times 60 \times 660)$
Distance from earth to sun	93,312,000 miles
(6 to the power 6x6,000)	

The sheer force of Michell's general theory – the framework surrounding these celestial distances whose measurements in miles factorize by 6 and 12 – is compelling. Consider, also, these additional facts that he deduced:

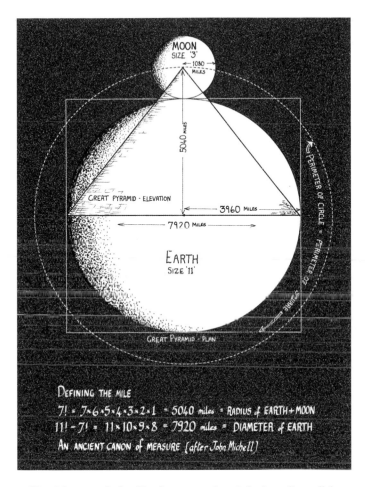

The Moon and the Earth square the circle in miles. John Michell's cosmological diagram, drawn by Robin Heath, (courtesy Wooden Books).

3,168ft = 1,164 megalithic yards of 2.722ft

31,680ft = 6 miles

31,680 inches = ¹/₂ a mile and 31,680 yards = 12 × 12 furlongs

31,680 furlongs = 3,960 miles = radius of the earth [and so on]

Adapted from an article contributed to The Yardstick, the BWMA journal, by Robin Heath, author of four books, including Sun, Moon and Stonehenge, and A Beginner's Guide to Stone Circles

Hodder, 1999

If one assumes that the megalithic yard represents a lunation period – the time between two new moons – then the length of 1ft (12") marks the required calendrical period between the end of the lunar year (which comprises 12 lunations taking 354.367 days) and the end of the solar year which comprises 365.242 days. Assuming that the MY was the primary unit, then the derivative foot appears to have formed a logical and essential part of the astronomical researches of our Neolithic ancestors. If, however, the foot *preceded* the MY (perhaps through the fact that 1,000th of 1 degree of arc around the equatorial circumference equals 365.221ft), then knowledge of the roundness of the earth must predate use of the MY – earlier than 3,000BC.

It would be logical for the key calendrical time interval – the 10.875 days elapsing between the ends of the lunar and solar years – to have been represented by an existing unit of length. It does appear that the foot of 12" was adopted accordingly.

My story doesn't end here, for there are 12.368 lunations in a year. Again assuming that 1 MY represents a lunation, then to tot up the exact number of lunations in the year, all one has to do is add one inch to a MY each and every time one observes a lunation. After the lunar year, the inches add up to 0.368 lunations, exactly the required over-run. When demonstrating accurate calendrical predictions with my student groups, I use a plastic Woolworth's foot-ruler, marked in inches, to predict lunations and eclipses precisely to the day, years in advance. Not very megalithic, perhaps, but highly accurate and apparently using those same measures of antiquity now rendered illegal by the government of the same lands that originally built those great stone circles.

Many of the greatest metrologists have suggested that the

primary units of length were derived from units of time. They all concluded that such measures were so strictly defined and so rigorously organized that they must have a basis on some absolute natural measure. I submit that this 'natural measure' was the lunation period. But whatever *length* a system of weights and measures was based upon, there was a coordinating link to units of *time*. The road to effective weights and measures begins with time measurement, leads to length, and finally to areas, volumes and weights. Accounts of the MY totally confirm this thinking; the cycles of the sun and moon defining the *absolute natural measure*.

So the MY may be considered a calendrical analogue of the lunation period, the foot (and the cubit*) being proportioned within it to reveal the length of the solar year, following the twelfth lunation. The accuracy is astonishing, to within less than half an hour in one solar year. The conclusions from this research have enormous implications for archaeology and the understanding of human prehistory. They confirm the primary connection between mensuration and the moon.

They also clearly offer one plausible explanation for the historical basis of the foot and its twelve-fold division into inches.

*Robin Heath provides further corroboration by showing that 1 Megalithic Yard (2.72ft) *exactly equals* 1 ft *plus* 1 ancient Egyptian 'Royal Cubit' of 1.72ft, but we cannot expand upon that here.

Temperature of a few of the principal Hot Springs.

Matlock	66°	Borset	132°
Bristol	74°	Aix	143°
Buxton	82°	Carlsbad	165°
Bath	114°	The Geysers, Iceland	212°
Berege	120°		

WEIGHTS AND MEASURES.

1.—MEASURE OF LENGTH.

12 Inches	1 Foot	40 Poles	1 Furlong
3 Feet	1 Yard	8 Furlongs	1 Mile
$5\frac{1}{2}$ Yards	1 Rod or Pole	$69\frac{1}{13}$ Miles	1 Degree

PARTICULAR MEASURES OF LENGTH.

$2\frac{1}{4}$ Inches	1 Nail	Used for Measuring Cloth.	4 Inches	1 Hand	
4 Nails	1 Quarter		6 Feet	1 Fathom	
4 Quarters	1 Yard		7·92 Inches	1 Link	For Mea. Land
5 Quarters	1 Ell		100 Links	1 Chain	

2.—MEASURE OF SURFACE.

144 Sq. Inches	1 Sq. Foot	40 Perches	1 Rood
9 Sq. Feet	1 Sq. Yard	4 Roods	1 Acre=10 Sq. Chains
$30\frac{1}{4}$ Sq. Yards	1 Perch or Rod	640 Acres	1 Sq. Mile

3.—MEASURE OF SOLIDITY AND CAPACITY.

Division I, Solidity.

1,728 Cubic Inc. ..1 Cubic Foot | 27 Cubic Feet ....1 Cubic Yard

Division II.

Imperial Measures of Capacity for all liquids, and for all dry goods except such as are comprised in the third division.

		Cubic Inches.		C. In.	
4 Gills	1 Pint	34·65925	or	$34\frac{2}{3}$	*nearly*
2 Pints	1 Quart	69·3185	„	$69\frac{1}{3}$	„
4 Quarts	1 Gallon	277·274	„	$277\frac{1}{4}$	„
2 Gallons	1 Peck	554·548	„	$554\frac{1}{2}$	„
8 Gallons	1 Bushel	2218·192	„	$2218\frac{1}{5}$	„
		Cubic Feet.		C. Ft.	
8 Bushels	1 Quarter	10·26936	or	$10\frac{1}{4}$	*nearly*
5 Quarters	1 Load	51·34681	„	$51\frac{1}{3}$	„

Division III.

Imperial measure of Capacity for coals, culm, lime, fish, potatoes, fruit, and other goods commonly sold by *heaped measure*:—

2 Gallons	1 Peck	$704\frac{1}{2}$	Cubic Inches, *nearly*
8 Gallons	1 Bushel	2815	
3 Bushels	1 Sack	$4\frac{8}{9}$	Cubic Feet, *nearly*
12 Sacks	1 Chaldron	$58\frac{2}{3}$	

The Imperial Gallon contains exactly 10 lbs Avoirdupois of pure water; consequently the Pint will hold $1\frac{1}{4}$-lb., and the Bushel 80-lbs.

4.—MEASURE OF WEIGHT.

Division I, Avoirdupois Weight.

$27\frac{11}{32}$ Grains	1 Dram	$27\frac{11}{32}$	Grains
16 Drams	1 Ounce	$437\frac{1}{2}$	„
16 Ounces	1 Pound (lb.)	7000	„
28 Pounds	1 Quarter (qr.)		
4 Quarters	1 Hundred Weight (Cwt.)		
20 Cwt.	1 Ton		

This weight is used in almost all commercial transactions, and in the common dealings of life.

Particular weights belonging to this division:—

		Cwt.	qr.	lb.	
8 Pounds	1 Stone				Used for Meat.
14 Pounds	1 Stone	0	0	14	
2 Stones	1 Tod	0	1	0	
6½ Tods	1 Wey	1	2	14	Used in the Wool Trade.
2 Weys........	1 Sack	3	1	0	
12 Sacks........	1 Last	39	0	0	

DIVISION II, TROY WEIGHT.

24 Grains........	1 Pennyweight........	24	
20 Pennyweights..	1 Ounce	480	Grains.
12 Ounces	1 Pound	5760	

TROY WEIGHT is used for weighing gold, silver, and precious stones except diamonds. The *carat* = 3⅙ grains, used for weighing diamonds, is divided into *halves, quarters,* &c. By a different division and sub-division of the ounce, it is also used in the compounding of medicines, and is then called—

APOTHECARIES' WEIGHT.

20 Grains	1 Scruple	8 Drams	1 Ounce
3 Scruples	1 Dram	12 Ounces	1 Pound

5.—MEASURE OF TIME.

60 Seconds	1 Minute
60 Minutes	1 Hour
24 Hours	1 Day
7 Days	1 Week
28 Days	1 Lunar Month
28, 29, 30, or 31 Days	1 Calendar Month
12 Calendar Months	1 Year
365 Days	1 Common Year
366 Days	1 Leap Year

6.—ANGULAR MEASURE, or DIVISIONS OF THE CIRCLE.

60 Seconds	1 Minute
60 Minutes	1 Degree
30 Degrees	1 Sign
90 Degrees	1 Quadrant
360 Degrees	1 Circumference

FRENCH AND ENGLISH MEASURES AND WEIGHTS COMPARED.

MEASURES OF LENGTH.

French.	English.	English.	French.
1 Millimetre	0·03937 Inch.	1 Inch ..	2·539954 Centimetres
1 Centimetre....	0·393708 „	1 Foot ..	3·0479449 Decimetres
1 Decimetre....	3·937079 „	Yard Imp..	0·91438348 Metre
1 Metre	39·37079 „	Pole or Perch..	5·09211 Metres
Myriametre	6·21382 Miles	Mile........	1609·3149 Metres

SQUARE MEASURES.

French.	English.	English.	French.
1 Metre sq....	1·196033 Yd. sq.	1 Yd. sq. ...	0.836097 Metre sq.
1 Are	0·098845 Rood	1 Rod ....25·291939 Metres sq.	
1 Hectare	2·473614 Acres	1 Acre ...	0·404671 Hectares

SOLID MEASURES.

French.	English.	English.	French.
1 Litre......	{ 1·760773 Pint { 0.2200967 Gal.	1 Pint	0·567932 Litre
		1 Quart	1·135864 Litre
1 Decalitre	0·2200967 Gals.	1 Gallon Imp.	4·84345794 Litres
1 Hectolitre ..	22·009667 Gals.	1 Quarter ..	2·907813 Hectolitres

Two pages from Harwood's Diary, an Almanack of 1864 published in London.

EPILOGUE

– the metric replacement?

1. A metric system, in which all measurements should be multiplied or divided by ten, was first devised by a priest of Lyons, Gabriel Mouton (which aptly means 'sheep') in 1670, although his idea was to take a one-thousandth part of the nautical mile (about 6 ft) as the unit of length. The *metre* was introduced by French revolutionaries in 1795, calculated (inaccurately) as a ten-millionth part of the distance of a line running on the Earth's surface through Paris, from the North Pole to the Equator. The relevance of that line to human measurement or practical purposes remains unclear, except for the kilometre to replace the nautical mile; for the French were envious of British maritime supremacy and had wanted the prime meridian of longitude to run through their capital rather than through Greenwich. This far-fetched formula produced an arbitrary unit which was, very neatly, just larger than the detested English yard.

2. A standard metre bar, made of platinum-iridium alloy, was kept in Sèvres, a few miles SW of Paris. The length was defined in 1960 in terms of optical wavelengths in a vacuum of radiation from atoms of krypton-86; but even that proved too simple and unreliable, so it was next expressed by reference to the wavelength of the radiation from an iodine-stabilized helium-neon laser, before finally resorting to measurement in terms of the speed of light; but the scientists must not be blamed for trying to improve the system!

3. From the metre as a unit of length, the French developed the *are* as the unit of area (i.e. 100 square metres or 1 square decametre), the *stere* as the unit of cubic measure (i.e. 1 cubic metre), the litre as the unit of capacity ($\frac{1}{1,000}$th of a *stere)* and the gram as the unit of mass of one cubic centimetre ($\frac{1}{1,000}$th of a

litre) of water. [see Appendix V] As the French tended to use centimetres rather than metres for everyday purposes, and also retained the second as the standard measure of time (even though the clock and calendar are wholly non-decimal), this spectrum of units became known as the centimetre-gram-second or CGS system, which was actually the brainchild of Lord Kelvin in 1873. But by the beginning of the 20th century the metre and kilogram (the weight of one litre of water) had become the inter-national metric units of length and mass respectively, and so the MKS system (originally proposed as early as 1901) was eventually adopted by the Conférence Générale des Poids et Mesures shortly after the second World War.

In 1950 the ampère (measuring electric current) was recognized as a fourth basic unit, then in 1960 at the 11th CGPM the kelvin (temperature) was included (273.15K = 0°C = 32°F) as well as the candela (luminous intensity), radian and steradian to constitute the Système International d'Unités (SI), which was adopted by the 11th CGPM in 1960 and further extended in 1964 at the 12th CGPM to include radionuclides, and yet again at the 14th in 1971 when the seventh base unit, the *mole,* was adopted.

4. How strange that Britain and America have managed to muddle along, ever since standardization of the Imperial system in 1825, without holding inter-national conferences at all! The result of these frequent metric conferences, in contrast, so far from producing one universal system, has been to leave many metric countries behind at different stages in this process of forced evolution, and for most countries to develop hybrid units which amalgamate metric and their own indigenous measures, adding further to the complexity and confusion of the global system. In Britain we already have the 'metric foot' of 30cm, equal to 11.81 inches.

The House of Commons recently produced its own style of ruler. A handsome design, emblazoned on one side with Parliament's logo (aptly the 'Traitor's Gate'), it measures 30cm in length, each side marked identically in cm-mm divisions, with the back completely blank. That's all the metric system can offer; one unit divided by ten and ten again. The official responsible was asked two questions. First, how can 30cm (which itself is not an authorized unit of length) be divided into 1 metre, which is the foundation stone of the entire metric structure? Answer: 'Sorry, it can't – other than 3.333 recurring!' Second, why 30cm anyway? Answer: 'It was as close as we could get to a foot.' How's that for shooting oneself in the foot? (30cm = 11.81 inches.) Our metric rulers can't even produce a metric ruler!

In contrast, the rulers produced by the British Weights and Measures Association show, on one edge, inches divided into $^1/_3$, $^1/_6$, $^1/_{12}$ and $^1/_{24}$, another edge $^1/_2$, $^1/_4$, $^1/_8$, $^1/_{16}$ and $^1/_{32}$, and on the other side $^1/_{10}$ and $^1/_{20}$ on one edge – for these are just additional fractions – and on the fourth edge the popular mapping scale of $1^1/_4$ inch to the mile, which is very close to 1:50,000 natural. Inherent complexity and confusion throughout the metric system arises from the distinction between the use of Latin prefixes for sub-units (decimetre, centimetre, etc) and Greek prefixes for multiple units (decametre, kilometre, etc). Indeed, there are 16 prefixes alone (8 Latin and 8 Greek) and symbols for each one: some in lower case and others in upper – so that if a lower case 'm' for 'milli' is used instead of 'M' for 'mega', the error could be by a factor of a thousand million.

5. To quote Derek Turner (Editor of *Right Now!*): "Technocrats are not susceptible to the visceral urges felt by us lesser mortals. Not for them irrational appeals to tradition, nation, family, love, honour, property, competition, music or metaphysics. They are very much like their spiritual predecessors in Revolutionary France, who devised the metric system in the interests of greater efficiency and as a means of divorcing the French from their glorious past. There are discernible connections between the clicking of slide-rules and the rattling of tumbrils. The American writer Bill Kauffman, in one of his superb essays, shows that he

understands the links between the phenomena of internationalism and metrification. Among the unsung patriots of our day, my everyday heroes, are the ornery old men who speak of quarts, not litres, refractory kids who flunk tests on their metric conversion tables, and the track officials who still stage 100-yard dashes and mile runs. Nothing about global markets, global culture or global government is inevitable in a world where the number of independent nations has increased five-fold since 1950."

6. Furthermore, the United Nations publishes a 138-page handbook, World Weights and Measures, detailing national currency and measurement systems. This volume destroys the myth of global metrication. It reveals not only the surprising number of countries still using the pint-foot-pound system but also how widespread, within countries that have long been officially metric, is the continuing use, for everyday purposes, of their indigenous units. The book recognizes that: "It would not be practical to show, for each of the thousands of units currently in use, equivalents in all the other systems of measurement." Indeed, it lists about 1,675 non-metric units, their values and countries of origin. So, ironically, it is the monolithic UN that provides the best possible proof that the metric system has a far from firm grip on the world.

7. Surely the most stupid and perverse example of metrication was by Radio 4, announcing the news of the superb achievement by the British *Thrust SSC* team in raising the world land-speed record past Mach 1 – on the sands of the Nevada Desert – to "1,149 kilometres per hour" instead of the actual 714mph. Never mind that in Britain it remains legal to use 'mph' as we continue to do universally; never mind that in the USA 'kph' is meaningless: here we have the BBC, which proclaims its commitment to education, referring to a speed in 'kph' over the prescribed distance of the *measured mile!*

8. To quote John Strange of BWMA: "The measurement of angles, by the arcs they cut on a circle, is as old as the notion of angle itself, and was already known to the Babylonians 4,000

years ago*. They divided the circle into 360° and introduced the sexagesimal scale. 60 was probably chosen because it is divisible by 2, 3, 4, 5 and 6 (as well as 10, 12, 15, 20 and 30). So when the Greeks required a finer division of the circle, they naturally divided each degree into 60 minutes. We still use these units today, despite French efforts to decimalize angles. (The French tried dividing the right angle into 100 grades or gons but the system failed because the sexagesimal system was so well rooted, both for geometry and for measuring time). *[*Megalithic Britons knew it, too.*]

Two right angles equal 200 *gons*, 180° or *pi* (3.1416) radians. There is, however, a technical difficulty. As the Greeks recognized, angles are not true magnitudes and the methods developed in Book V of Euclid's *Elements* cannot be applied to them. Suppose for a moment that the Earth were a perfect sphere. The meridians are semi-circles, drawn on its surface, whose end points are the North and South Poles; the polar axis is the straight line which runs through the centre of the Earth, joining both poles. Then the nautical mile is the distance between two points on the same meridian whose latitudes differ by one minute.

So the distance along a meridian from either of the poles to the equator must be 90 times 60 or 5,400 nautical miles; but because the Earth is squashed at the poles, its curvature is greater near the equator than elsewhere and consequently the nautical mile, as defined above, varies from nearly 6,046 ft near the equator to about 6,107 near the poles. The Admiralty decided that it should be fixed at 6,080 ft, but the international nautical mile has since been defined as 1.852 km, which is just under 6,076' 1½". The French wanted to define the metre in terms of the distance from the North Pole to the equator measured along the Paris meridian. This distance was to be 10 megametres so that any two points on the meridian, whose latitude differed by one grade, would be 100 km apart. Each grade is divided into 100 parts and the kilometre was to replace the nautical mile. But the Earth is divided by meridians into 24 time zones. The two meridians which bound one of these time zones meet at the poles at an angle of 15° (360°/24). This relationship between angles and time zones is rather awkward if the angles are measured in grades. The French had foreseen the difficulty and, on 5 October 1793, the Convention decreed the decimalization of time. This proved a failure and by a further decree on 7 April 1795, the date of birth of the metric system, suspended the operation of the earlier one indefinitely. So, in a temporal sense, the metric system was still-born. As a result, traditional units for time are still used universally and navigators still use nautical miles.

Scientists, who do not mind if the unit they are using is sometimes disproportionate to the thing they are measuring, are quite happy to use the *second* for time and the *metre* for distance."

9. From *Men and Measures* by Edward Nicholson (1912): "In England a few genuine enthusiasts, and many more who have caught the scientific and cosmopolitan craze, take to the metric system as they take to learning Esperanto, and so long as they have not to use the one for business or the other in conversation, their enthusiasm lasts, especially when it affords opportunities for showing themselves friends of science and progress. But when the contagion spreads so wide that it threatens to revolutionize the habits and customs of a nation and its whole manufactures and trade, the danger is most serious.

The favour which the metric system has found amongst a small proportion of English people is largely due to their ignorance of their own system, an ignorance very excusable when there exists no official statement of our system, or even of its standards."

10. To quote Professor C Piazzi Smyth (former Astronomer Royal for Scotland) in 1872: "We, as a nation, have done as yet little beyond merely offering a dull resistance to any kind of change in things which we undervalued, but to which we have

clung with the inertia of custom and wont, because they had come into our possession we knew not how or when."

11. The British Standards Institute has adopted 30cm as a building module, following the disuse of the decimetre; but now even the cm is slowly dying, to leave only the metre and millimetre. [So BBC weather presenters call 24 inches of rain, not 61cm but 610mm, which is not only absurdly inflated but also absurdly precise!] As it happens (as Arthur Whillock has pointed out), 30cm is very close to the original Greek Attic foot of sixteen digits; enlarged from the Pythic foot of 12 digits [see para. 4 above]. Also, it is worth noting that the only point of coincidence between centimetres and inches is the exact equivalence of 127cm and 50". Incidentally, the equivalent of 1in is always given as 2.54cm but actually it is 2.5399978cm. Whillock commented (1973): "This new metric foot, with its twelve 'inches' of 25mm (used for timber sizes, bolt lengths, tape widths, etc), has caused alarm in some quarters, but give it time; the metre may become the new 'ell' of 40 inches! Relieved of the need to support technical work for the time being, our foot will be free to adjust itself to the value used by the designer of the Parthenon, and it is fitting that his descendants are engaged in putting measurement back onto the rails it left some 150 years ago."

12. To quote again from Jacob Bronowski's *The Ascent of Man*: "Take a beautiful cube of pyrites. Or to me the most exquisite crystal of all, fluorite, an octahedron. (It is also the natural shape of the diamond crystal.) Their symmetries are imposed on them by the nature of the space we live in – the three dimensions, the flatness in which we live. And no assembly of atoms can break that crucial law of nature. Like the units that compose a pattern, the atoms in a crystal are stacked in all directions. So a crystal, like a pattern, must have a shape that could extend or repeat itself indefinitely. That is why the faces of a crystal can only have certain shapes; they could not have anything but the symmetries in the pattern. For example, the only rotations that are possible

go twice or four times for a full turn, or three times or six times – not more. And not five times. You cannot make an assembly of atoms to make triangles which fit into space regularly five at a time."

13. To quote from an article in *Nature* published in 1922: "In view of the vigorous and sustained efforts of the exponents of the metric system, and the eminent names that are to be found among them, it is perhaps not a little surprising that it makes so little progress towards general acceptance in Great Britain. The Weights and Measures Act of 1897 legalized the metric denominations for use in trade, and was expected to lead to its advantage being so generally recognized that the Imperial system would soon disappear.

Twenty-five years have now elapsed and the position is almost unchanged. In fact, the policy of compulsory introduction of the metric system by law, which formerly was always strongly supported, was ruled out by the Metric Committee of the Conjoint Board of Scientific Societies in its Report of 1919...The subject of compulsion is not likely again to be seriously considered for some time at least."

14. To quote again from John Neal's *Opus 2 – All done with Mirrors:* "The universal acceptance of the metric system was assured by virtue of its convenience to the international banking regime, all other considerations being subservient to its requirements. The primary reason that the United States has resisted conversion for so long is that its monetary system was decimalised from the very foundations of the Nation.

With its international currency trading already conformed, no necessity was seen to interfere with its internal standards. Modern money itself being artificial, it is perfectly suited to an artificial counting method that can only express quantity; whereas in the fields of the arts and sciences, a system of harmonic proportions among quantitative expressions is ideally satisfied by the traditional measures.

The metre, when legalised by the British in 1864 for use in contracts, was resolved to be 39.3708 inches. Yet when legalized in 1897 for purposes of general trade, it was more accurately standardized as 39.370113 inches. But when formally defined for recognition by the United States in 1866, there it was 39.37 inches. Again, when the metre was adopted in Japan in terms of the *shaku*, the definition expressed is also minutely at variance with both the British and US standards. Each nation has a slightly different interpretation. This lack of agreement on conversion rates can have disastrous consequences for scientific endeavours involving time and distance." To summarize: the metric system was devised by scientists for scientists, whose purposes it serves very well; but that doesn't mean that it's appropriate for everyday use.

15. To quote again from Martin Doutre's *Ancient Weights, Volumes and Measurements*: "The ancient universal metrological science, of which the British Imperial standard is a 'walking wounded' surviving remnant, was based upon the true size of the Earth, the precise cycles of the sun and moon and incremental values that worked wonderfully well with *pi* for successful world navigation. The system was so advanced mathematically that it tamed for use the difficult *phi* formula and applied it in many building applications. This old standard was a brilliant, precise and all-encompassing, scientific construct that addressed every arising calculation need confronted by ancient civilizations. The system is as relevant today as it was in antiquity and has never been superseded.

The metrication initiative is to metrology what Esperanto is to languages. The metric system is an incomplete entity and could not have subsisted by itself in antiquity. The ancient civilizations used the whole range of numbers from 1 to 13 and increases thereof. The metric system, most certainly, existed in antiquity, alongside all the other number systems. Ancient societies would have viewed it as sheer madness to discard over 90% of the repertoire. The Egyptians counted in base 10 but they used '6, 7 and 11' numerical families as well. We do need the convenience of a metric system, within a wider metrological system, for some limited, decimalised functions; but to lose our grasp on the old, universal metrological system is to lose the only key to unlocking the mysteries of the past."

APPENDICES

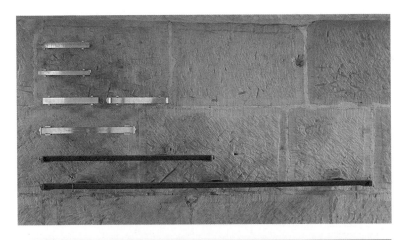

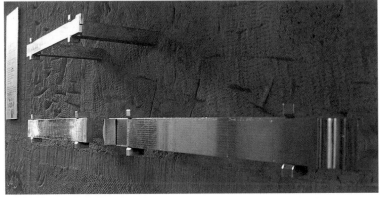

On the inside wall of an arch over the pavement which forms the base of an early 16th century clock tower that marks the official centre of Berne, Switzerland's ancient capital, are mounted 6 horizontal bars displaying various linear units, including the Berne cubit and conversions from the local 'shoe' to the Swiss foot -- in the mediaeval canton of Berne, 'shoe' meant a foot length. Photographs by Mairi Rivers.

Appendix I

Foreign equivalents of the Foot and Continental Mile measurements

Several foreign units of linear measure correspond to the Imperial foot. In most cases, the unit divides into 12ths, like the inch, or *multiplies* by 12 into a much larger unit: e.g. the Netherlands *Roede* = 12 *Voet* (each equalling 12.06 inch); like the Austrian *Rute* = 12 Fuss (see below).

Country	Name	Length in inches
Turkey	¹/₂ pik	13.5
Portugal	pe	12.96
France	pied	12.79
Austria	fuss	12.44
Germany	fuss	12.36
Norway & Denmark	fod	12.36
Babylonia (ancient)	(?)	12.24
Greece	pous	12.08
Russia	foute	12.00
Japan	shaku	11.93
Belgium	pied	11.81
Roman (ancient)	pes	11.60
Holland	voet	11.15
Spain	pie	10.98

Continental Mile Measurements

Austrian mile	8,296 yards
Spanish mile	5,028 yards
Russian verst	1,167 yards
Italian mile	1,467 yards
Polish mile	4,400 yards

Appendix II

Some quotations from Shakespeare

contributed by John Constable

Faith, here's an equivocator, that could swear in both the scales
against either scale   *Macbeth: Act II Scene iii line 9*

Now would I give a thousand furlongs of sea for an acre of
barren ground   *The Tempest: I i 70*

With one soft kiss a thousand furlongs ere.
With spur we heat an acre   *Winter's Tale: I ii 94*

My bosky acres, and my unshrubbed down *The Tempest: IV i 81*

Between the acres of the rye.
With a hey, and a ho, and hey-nonny-no *As You Like It: V iii 24*

Over whose acres walked those blessed feet.
Which fourteen hundred years ago were nailed,
For our advantage, on the bitter cross   *1 Henry IV: I i25*

Now get you to my lady's chamber, and tell her, let her paint
an inch thick, to this favour she must come   *Hamlet V: I 212*

And if thou prate of mountains, let them throw
Millions of acres on us   *Hamlet: V i 302*

O coz, coz, coz, my pretty little coz,
that thou didst know how many fathom deep I am in love!

As You Like It: IV i 218

Full fathom five thy father lies   *The Tempest: I ii 394*

Where fathom-line could never touch the ground

1 Henry IV: I iii 204

Another of his fathom they have none   *Othello: I i 153*

And not the worst of the three but jumps twelve foot and a
half by the squier [square]   *The Winter's Tale: IV iii 349*

Three foot of it doth hold: bad world the while!

King John: IV ii 99

If I travel but four foot by the square further afoot,
I shall break my wind   *1 Henry IV: II ii 13*

By the good gods, I'd with thee every foot   *Coriolanus: IV i 56*

Item: sack, two gallons   *1 Henry IV: II iv 595*

That you should have an inch of any ground
To build a grief on   *2 Henry IV: IV i 109*

I'll show thee every fertile inch o'the island

The Tempest: II ii 160

I will fetch you a tooth-picker now from the furthest
inch of Asia   *Much Ado: II i 277*

I'll not budge an inch, boy   *Taming of the Shrew: Induction. i. 14*

One inch of delay more is a South Sea of discovery

As You Like It: III. ii. 207

For every inch of woman in the world,
Ay, every dram of woman's flesh is false

The Winter's Tale: II. i. 136

I have speeded hither with the very extremist inch of possibility

2 Henry IV: IV. iii. 38

Being now awake, I'll queen it no inch further

The Winter's Tale: IV. iii. 462

My inch of taper will be burnt and done   *Richard II: I. iii. 223*

Not an inch further   *1 Henry IV: II. iii. 119*

Ay, every inch a king   *King Lear: IV. vi. 110*

Am I not an inch of fortune better than she?

Antony and Cleopatra: I. ii. 61

For I'll cut my green coat, a foot above my knee,
And I'll clip my yellow locks, an inch below mine eye

The Two Noble Kinsmen [authorship disputed]

Whom I, with this obedient steel – three inches of it –
Can lay to bed for ever   *The Tempest: II i 291*

Ask them how many inches is in one mile: if they have
measur'd man. The measure then of one is easily told.

Love's Labour's Lost: V ii 189

Am I but three inches?
Why, thy horn is a foot; and so long am I at the least.

Taming of the Shrew: IV i 29

Eight yards of uneven ground is threescore and ten miles afoot
with me   *1 Henry IV: II ii 27*

And buckle in a waist most fathomless
With spans and inches so diminutive

Troilus and Cressida: II ii30

As many inches as you have oceans   *Cymbeline: I ii 21*

My sweet ounce of man's flesh!   *Love's Labour's Lost: III i 142*

Nay then, I must have an ounce or two of this malapert blood
from you   *Twelfth Night: IV i 48*

Give me an ounce of civet, good apothecary, to sweeten my
imagination   *King Lear: IV vi 132*

Weigh you the worth and honour of a king / So great as our
dread father in a scale / Of common ounces?
                                              Troilus and Cressida: II ii 26

Good faith, a little one; not past a pint, as I am a soldier
                                              Othello: II iii 69

The pound of flesh which I demand of him
                                              Merchant of Venice: IV i 99

Three pound of sugar; five pound of currants…four pound of
prunes   *The Winter's Tale: IV ii 40*

Butter at eleven pence a pound
                                              Sir Thomas More [authorship disputed]

Indeed, I am in the waist two yards about
                                              Merry Wives of Windsor: I iii 43

I looked a' should have sent me two-and-twenty yards of satin
                                              2 Henry IV: I ii 48

Appendix III

from the REPORT from the SELECT COMMITTEE
on WEIGHTS AND MEASURES
communicated by the Commons to the Lords, 1816

Your Committee, in the first place, proceeded to enquire what measures had been taken to establish uniform Weights and Measures throughout the Kingdom. They found that this subject had engaged the attention of Parliament at a very early period. The Statute Book, from the time of Henry the Third, abounds with Acts of Parliament enacting and declaring that there should be one uniform Weight and Measure throughout the Realm; and every Act complains that the preceding Statutes had been ineffectual, and that the Laws were disobeyed.

In order to obtain some information as to what were the best means of comparing the standards of length, with some invariable natural standard, your committee proceeded to examine.....From the evidence of these gentlemen, it appears that the length of a pendulum making a certain number of vibrations in a given portion of time, will always be the same in the same latitude; and that the standard English yard has been accurately compared with the length of the pendulum, which vibrates 60 times in a minute in the latitude of London. The length of this pendulum is 39.13047 inches, of which the yard contains 36. Any expert watch-maker can easily adjust a pendulum, which shall vibrate exactly 60 times a minute.

The French Government have adopted as the standard of their measures, a portion of an arc of the meridian, which was accurately measured. The standard metre, the 10,000,000th part of the quadrant of the meridian, is engraved on the Platina Scale preserved in the National Institute. It has been compared with the English standard yard....and was found to exceed it, at the temperature of 32 degrees, by 3.3702 inches; and at the temperature of 55 degrees, by 3.3828 inches. The standard yard may therefore be at any time ascertained, by a comparison either with an arc of the meridian, or the length of a pendulum, both of which may be considered as invariable.

The standard of linear Measure being thus established and ascertained, the measures of capacity are easily deduced from it, by determining the number of cubical inches which they should contain. The standard of weight must be derived from the measures of capacity, by ascertaining the weight of a given bulk of some substance of which the specific gravity is invariable. Fortunately that substance which is most generally diffused over the world, answers this condition. The specific gravity of pure water has been found to be invariable at the same temperature; and by a very remarkable coincidence, a cubic foot of pure water (or 1,728 cubical inches) at the temperature of $56^1/_2$ degrees by Fahrenheit's thermometer, has been ascertained to weigh exactly 1,000 ounces Avoirdupois, and therefore the weight of 27.648 inches is equal to one pound Avoirdupois.

This circumstance forms the groundwork of all succeeding observations of Your Committee. Although in theory the standard of weight is derived from the measures of capacity, yet in practice it will be found more convenient to reverse this order. The weight of water contained in any vessel, affords the best measure of its capacity, and is more easily ascertained than the number of cubical inches by gauging. Your Committee therefore recommends that the measures of capacity should be ascertained by the weight of pure or distilled water contained by them, rather than by the number of cubical inches, as recommended in the 4th Resolution of the Committee of 1758.

Your Committee is also of opinion, that the standard gallon, from which all the other measures of capacity should be derived, should be made of such a size as to contain such a weight of pure water of the temperature of $56^1/_2$ degrees, as should be expressed in a whole number of pounds Avoirdupois, and such also as would admit of the quart and pint containing integer numbers of ounces, without any fractional parts. If the gallon is made to contain 10 pounds of water, the quart will contain 40 ounces and the pint 20. If this gallon is adopted, the bushel will contain 80 lbs of water, or 2,211.84 cubical inches; the quart 69.12 cubical inches... the pint 34.56 cubical inches (which is exactly $\frac{1}{50}$th part of a cubical foot) ...

from the FIRST REPORT of the COMMISSIONERS appointed to consider the subject of Weights and Measures 1819

The Subdivisions of Weights and Measures, at present employed in this Country, appear to be far more convenient for practical purposes than the Decimal Scale, which might perhaps be preferred by some persons, for making calculations with quantities already determined. But the power of expressing a third, a fourth, and a sixth of a foot in inches, without a fraction, is a peculiar advantage of the Duodecimal Scale; and for the operations of weighing and of measuring capacities, the continual division by Two renders it practicable to make up any given quantity, with the smallest possible number of standard Weights or Measures, and is far preferable, in this respect, to any decimal scale. We would therefore recommend, that all the multiples and subdivisions of the Standard to be adopted should retain the same relative proportions to each other, as are at present in general use.

from the THIRD REPORT of the COMMISSIONERS' 1821

...And we have found by the computations...that the weight of a cubic inch of distilled water, at 62 degrees Fahrenheit, is 252.72 grains of the Parliamentary Standard Pound of 1758, supposing it to be weighed in a vacuum. We beg leave therefore finally to recommend, with all humility, to Your MAJESTY, the adoption of the Regulations and Modifications suggested in our former Reports; which are principally these:

That the Parliamentary Standard Yard, made by Bird in 1760, be henceforwards considered as the authentic legal Standard of the British Empire; and that it be identified by declaring, that 39.1393 inches of this standard [of which the yard contains 36] at the temperature of 62 degrees Fahrenheit, have been found equal to the length of a Pendulum supposed to vibrate seconds in London, on the level of the sea, and in a vacuum.

That the Parliamentary Standard Troy Pound, according to the two pound weight made in 1758, remain unaltered; and that 7,000 Troy Grains be declared to constitute an Avoirdupois Pound; the cubic inch of distilled water being found to weigh at 62 degrees, in a vacuum, 252.72 Parliamentary grains.

That the Ale and Corn Gallon be restored to their original equality, by taking, for the statutable common Gallon of the British Empire, a mean value: such that a gallon of common water may weigh 10 pounds avoirdupois in ordinary circumstances, its content being nearly 277.3 cubic inches; and that correct Standards of this IMPERIAL GALLON, and of the Bushel, Peck, Quart and Pint, derived from it, and of their parts, be procured without delay for the Exchequer, and for such other offices in Your Majesty's dominions, as may be judged most convenient for the ready use of Your Majesty's subjects.

Appendix **IV**

from the Report submitted to the US Congress in 1821 by JOHN QUINCY ADAMS, then Secretary of State, prior to his election as President in 1825. He had been commissioned to produce this Report in order to help Congress determine the system of weights and measures they should adopt for the new Nation. It is a most impressive document, reflecting those same fundamental truths concerning the origins and current condition of the imperial system, and also the deficiencies of the metric, that still apply today.

Considered as a whole, the established weights and measures of England are but the ruins of a system, the decays of which have been often repaired with materials adapted neither to the proportion nor to the principles of the original construction. The metrology of France is a new and complicated machine, formed upon principles of mathematical precision, the adaptation of which to the uses for which it was devised is yet problematical, and abiding, with questionable success, the test of experiment.

To the English system, belong two different units of weight and two corresponding measures of capacity, the natural standard of which is the difference between the specific gravities of *wheat* and *wine*. To the French system, there is only one unit of weight and one measure of capacity, the natural standard of which is the specific gravity of water.

The French system has the advantage of unity in the weight and the measure, but has no common test of both: its measure gives the weight of water only. The English system has the inconvenience of two weights and two measures; but each measure is, at the same time, a weight. Thus the gallon of wheat and the gallon of wine, though of different dimensions, balance each other. A gallon of wheat and a gallon of wine each weigh eight pounds *avoir-dupois*.

The *litre* in the French system is a measure for all grains and all liquids; but its capacity gives a weight only for distilled water. As a measure of corn, of wine, or of oil, it gives the space they occupy, but not their *weight*. Now, as the weight of these articles is quite as important in the estimate of their quantities as the space which they fill, a system which has two standard units for measures of capacity, but of which each measure gives the same weight of the respective articles, is quite as uniform as that which, of any given article, requires two instruments to show its quantity – one to measure the space it fills and another for its weight. In the difference between the specific gravities of corn and wine, nature has also dictated two standard measures of capacity, each of them equiponderant to the same weight.

This diversity existing in nature, the Troy and Avoirdupois weights, and the corn and wine measures of the English system are founded upon it. In England it has existed as long as any recorded existence of man upon the island; but the system did not originate there. The weights and measures of Greece and Rome were founded upon it. The Romans had the *mina* and the *libra*, the nummulary pound [relating to coinage] of 12 ounces and the commercial pound of 16. The avoirdupois pound came through the Romans from the Greeks, and through them, in all probability, from Egypt. Of this there is internal evidence in the weights themselves, and in the remarkable coincidence between the cubic foot and the 1,000 ounces avoirdupois, and between the ounce avoirdupois and the Jewish *shekel;* and if the *shekel* of Abraham was the same as that of his descendants, the avoirdupois ounce may, like the cubit, have originated before the flood.

The result of these reflections is that the uniformity of nature for ascertaining the quantities of all substances, both by gravity and by occupied space, is a uniformity of *proportion*, and not of i*dentity;* that, instead of one weight and one measure, it requires two units of each, *proportioned* to each other; and that the original English system of metrology, possessing two such weights and two such measures, is better adapted to the only uniformity applicable to the subject, recognized by nature, than the new French system which, possessing only one weight and one measure of capacity, identifies weight and measure by only the single article of distilled water; the English uniformity being relative to the *things* weighed and measured, and the French only to the *instruments* used for weight and mensuration.

Later in this Report, Adams commented on the decimal principle. "It can be applied, only with many qualifications, to any general system of metrology; that its natural application is only to numbers; and that time, space, gravity and extension inflexibly reject its sway...

It is doubtful whether the advantage to be obtained by any attempt to apply decimal arithmetic to weights and measures, would ever compensate for the increase of diversity which is the unavoidable consequence of change. Decimal arithmetic is a contrivance of man for computing numbers, and not a property of time, space or matter. Nature has no partialities for the number ten; and the attempt to shackle her freedom with it will forever prove abortive."

The 'Union' bushel of 1707, issued to Scottish burghs to serve as the new legal standard. Courtesy of the National Museum of Scotland.

Appendix V

The Gallon

The gallon illustrates well the difference in outlook between the metric and imperial systems. Referring to Appendix III (particularly the 1821 Report and penultimate para. of the 1816 Report), we see the Commissioners' concern to provide a measure that was not only precise but also practical. So, if you want to find a container's capacity, all you have to do is to weigh it empty and then weigh it again filled with pure water. The difference between the two weights in ounces is the volume of the container in fluid ounces. As the gallon = 8 pints and each pint = 20 fl.oz, therefore 1 gallon weighs 160 fl.oz or 10 pounds. (For greater precision, the water should be at a temperature of 62°F.)

Let us look first at the French system. How were they to choose a unit of mass? They decided that the unit should be the mass of one cubic decimetre of water. But that mass depends on the temperature of the water – so that had to be decided upon. Now, if the temperature varies slightly from 39.16°F, the density of water hardly changes, so it makes sound theoretical sense to choose that point in order to minimize the effect of any error in temperature, which the French authoritiesaccordingly did. They made a cylinder of platinum-iridium whose mass was *very nearly* that of 1 cubic decimetre of water at its most dense...and that is the kilogram. When it was subsequently discovered that the standard kilogram was very slightly heavier than intended, the litre was defined as the volume occupied by 1 kilogram of water at its most dense.

This turns out to be very nearly 1,000.028 cubic centimetres – redefined in 1964 simply as 1,000 cu.cm. (The Revolutionaries should not be blamed for miscalculating the metre and the kilogram: As one of them said: *La République n'a pas besoin de savants*, so they guillotined Lavoisier, the discoverer of oxygen.)

Now it is a matter of common experience that if you venture a short way into the sea and pick up a stone, it feels heavier as soon as it is withdrawn from the water. When it's in the water, it is buoyed up. Archimedes' principle states that: When a solid body is immersed (wholly or only partly) in a fluid, it experiences an upward thrust equal to the weight of the fluid displaced.

Another obvious example is that of a balloon in air. Indeed, even a man experiences an upthrust from the surrounding air, but it's so small – about 3oz – as to be negligible. Now, *nos amis* did not overlook this effect when they weighed their water. So they ended up with a rather impractical definition of the litre. The water had to be about 39.16°F (pretty damn cold) and the weighing had to be performed in a vacuum!

Compare this with the imperial gallon. The temperature is a very reasonable 62°F, and we don't have to worry about the weight of the air displaced because that's taken care of. As the Third Report says – having spoken earlier of the theoretical weighing in a vacuum – "a gallon of common water may weigh ten pounds avoirdupois *in ordinary circumstances*" [my emphasis]. Does that not sum up our customary measures: 'common sense in ordinary circumstances'?

A gallon is today defined as 4.54609 litres – i.e. very nearly 277.4194328 cu.in. or 0.16054365323 cu.ft., while the density of water at 62°F is very nearly 997.68 oz per cu.ft. Consequently, the mass of 1 gallon of water at 62°F is very nearly $997.68 \times 0.16054365323 = 160.17119$ oz. But that gallon of water experiences an upthrust of 0.19515 oz, this amount being the weight of air it has displaced. So the gallon of water appears to weigh only 159.97604 oz. Yet that's not the end of the story, for the 10lb in the other scale pan has displaced 0.02393oz of air. So, finally, the gallon of water appears to weigh 159.99997oz – you could hardly ask for anything more accurate than that! Therefore, while the French method is theoretically capable of giving greater precision, it is not quite so accurate and less useful in practice, because we are not going to use water at 39.16°F and we are not going to weigh it in a vacuum.

Much of the data for this article are taken from The Weights and Measures Act of 1963

Appendix VI

Excerpts from the Rules of Lawn Tennis

The singles court is 78 ft long and 27 ft wide. The service court is 21 ft long. The doubles court is 36 ft wide. The net posts are 3 ft outside the court. The net is 3'6" high at the posts and 3'0" in the middle. The lines are between 1 and 2 inches wide except the base line which is between 2 and 4 inches wide.

The diameter of the balls is between two-and-a-half and two-and-five-eighths inches and their weight between 2 oz and 2 oz 1 dram. When dropped onto a concrete base from a height of 100 in., the ball shall bounce to a height between 53" and 58". The tests are performed at a temperature of 68°F and an atmospheric pressure of 29.95" of mercury."

Proof Spirit

In the old days, when a ship arrived from France with brandy on board, Excise men put some gunpowder on the quay-side, poured a little of the brandy over it and applied a match. If the fire fizzled out, the brandy was declared to be under proof; if it flared up it was over proof, but if it burned gently it was certified as proof spirit.

Nowadays, a volume of spirits is at proof (i.e. 100°) if it weighs $^{12}/_{13}$ of the weight of the same volume of water, both being at a temperature of 51°F. In practice, if we mix 2 pints 9fl.oz. (3lb 1oz.) of water with 3 pints (3lb) of pure alcohol we get about $5^{1}/_{4}$ pints of proof spirit; for there's some shrinkage when the liquids are mixed.

Spirits sold in the UK are usually 30 percent under proof (i.e. 70 degrees), corresponding to a 40 percent alcoholic content. This $^{7}/_{4}$ ratio reflects (approx.) the fact that pure alcohol equals $175^{1}/_{4}$ degrees proof ('75 over proof'). But the US proof scale differs: a US whisky of 80 degrees proof equating to 70 degrees on the UK scale.

Most Scotch whisky is exported at 86 Proof American while most American whiskeys are sold at 90 or 100 Proof American. Here is a table, comparing (in round figures) the UK and US standards, along with the metric index, which simply gives the actual percentages of pure alcohol (by *volume*, not by *mass:* drink-driving legislation refers to the *mass* of alcohol in a given *volume* of blood!).

UK Proof degrees	US Proof degrees	Metric % pure alcohol	
175	200	100	
100	114	57	
88	100	50	
85	98	49	
80	90	45	
75	86	43	
70	80	40	
65	74	37	
0	0	0	(water!)

Cheers!

Appendix VII

The King's Girth and the Cosmological Pattern of the Saxon Royal Court by John Michell

copyright reserved by the author, who generously gave BWMA permission for publication in our first edition

John Neal, John Michell's colleague, when sending us this article, commented: "The greatest metrologist of all, Livio Stecchini, found examples of a wide range of measurements engraved as standards on Roman monuments, in addition to those that we accept as 'Roman'. He stated, 'These monuments confirm that not only in Rome one used the natural version of the Roman foot, but also that in the ancient world all of the units of length formed a system and were used concurrently.' In addition, the practices of the Roman 'gromatici', or planners, who surveyed the towns and consecrated the surrounding area, would appear to be based on the identical canon as the Saxon therefore, our own. In spite of the fact that the Saxons preserved the Roman estate and land divisions, as in the 'ville' and the 'tun', there is no reason to believe that their methods were inherited from a Roman source....the only point on which I would disagree with John Michell's treatment of the evidence is his statement that the necessary knowledge is 'revealed' intermittently. My belief is that it has been continuously known from an immensely ancient source and has simply run down at different rates in different places. Although it is subject to sporadic renaissance, alas, the trend is ever downwards."

Athelstan, a grandson of King Alfred, was king of Mercia in 934 and the following year was elected 'king of all the English'. He codified the laws of the country and refined its standards of measure. One of these, known as the King's Girth, represented the distance from the monarch or his residence within which all offences were regarded as treason against the Crown. Its stated length was: 3 miles, 3 furlongs, 9 acres, 9 feet, 9 palms and 9 barleycorns. Of these units, the mile of 5,280 feet, the furlong of 660 feet and the foot itself are still in use today. The barleycorn was one third of an inch and the palm was 11 barley-corns or 3 $2/3^{rds}$ inches. The acre, which is now exclusively a unit of area, once had a linear application. According to Richard Bernese's work on land surveying published in 1537, its length was the shorter side of a rectangular strip measuring 4 by 40 perches.

The present accepted value of the perch is $16^{1}/_{2}$ feet, making a linear acre $4 \times 16^{1}/_{2}$ or 66ft [later known as a chain]. These were called field measures and were used for surveying arable land. Longer than them by 1 part in 11 were the units applied to forested land: the woodland perch of 18ft and its linear acre of 72ft. The linear acre in the King's Girth is this 72ft measure. The length of the King's Girth is therefore:

3 miles of 5,280ft	=	15,840 ft
3 furlongs of 660ft	=	1,980
9 linear acres of 72ft	=	648
9 feet	=	9
9 palms of 0.305556 ft	=	2.75
9 barleycorns of 0.027778ft	=	0.25

$$18,480.00\text{ft} = 3^{1}/_{2} \text{ miles } exactly!$$

Taking this measure as a radius, the area of the King's Girth is 7 miles wide. As a square it contains 49 square miles, or $38^{1}/_{2}$ if it is a circle.

This curiously detailed account of the various measures making up the King's Girth is also curiously imprecise. The Saxon lawmen make no mention of the shape of the area, whether it is a square or a circle, and we are not told whether the $3^{1}/_{2}$ miles is a radius or a diameter. These omissions are significant and imply that the King's Girth is not just a measured length or area but a symbol of an ideal state.

It is a symbol that relates to three different levels of reality. First, it is a cosmological pattern, expressing through number, measure and geometry, an aspect of the heavenly order. Coming down to earth, it delineates the ritual order of the Saxon court and state; and, on the material plane, it is the model of the ancient English manor system.

The basic pattern is neither a square nor a circle but both together – a circle contained by a square of 7 miles. The perimeter of the square is 28 miles, of the circle 22 miles. The square is divided into 49 smaller squares, each containing 1 square mile or 640 field acres of 43,560 square feet. This arrangement was exactly repeated in the USA, much of which was divided in the same way into sections of 640 acres.

As the ideal constitutional pattern of the royal court, the King's Girth diagram [see below] is divided into two main parts. The central square contains the royal park and residence, while the rest of the area stretches away for 3 miles on each side. This outer area is the royal domain or domestic hunting ground, well forested and stocked with game.

The central square, 1 mile wide, is divided into $8 \times 8 = 64$ squares, each 1 furlong wide. Sixty squares are occupied by parkland, while the central four squares contain the court precinct, the King's 24ft square court or chamber, the central platform of 6ft and the pole – 6 inches thick – that forms the axis of the whole ritual order.

Reflecting the pattern of the royal court, but on the less formal level of actual human life, is the manorial plan. The manor is a self-governing village whose main components correspond to each of the six units of measure in the King's Girth. The three smallest – the barleycorns, palms and feet – make up the central 24 ft-wide area representing the sanctuary that later became the site for a village church. The area measured by acres is the residential part of the manor; occupied by houses, workshops and gardens. Beyond that, measured by furlongs, are the cultivated fields, divided into long, parallel strips by the rod, chain and acre, and apportioned out each year between the manor families. Surrounding the fields is the greater part of the manor lands, stretching for 3 miles in each direction and consisting of meadows, commons and woodlands. These provide fuel, timber for building, pasturage for domestic animals, game and all other needs of the manor.

Here is a summary of the pattern in units of feet, with its symbolism on three levels: the heavenly, the kingly and the practical.

Unit	Length x 2	Total width	Functions and symbolism
9 barleycorns	0.5	0.5	world-axis, eternal law, central standard
9 palms	5.5	6	seat of divine justice, throne, central rock
9 feet	18	24	hall of the gods, king's court, central
9 acres	1,296	1,320	sanctum home of the gods, court precinct,
3 furlongs	3,960	5,280	manor dwellings, abode of the just, royal
3 miles	31,680	36,960	park, cultivated land, Elysian fields, royal forest, commons/woodland

This is a pattern of an ideal realm with a state constitution designed to reflect the order of Creation. It is not a contrived pattern, peculiar to the Saxons or any other race or age, but derives from an ancient tradition of spiritual and scientific knowledge that has left its mark in the cultures – and most plainly in the identical units of measure – of people in every continent. Hierarchical, cosmologically based societies in NW Europe, from the Bronze Age Celts to the mediaeval Vikings, have their prototype in this same numerical pattern which, according to classical historians, is revealed to humanity by the gods at different times and places. Everything in these societies was apparently centred upon the king who, like Athelstan, was elected as the best man among his peers.

But he was merely a symbol and executor of law, not the source of it. At the heart of all is a rock or altar with the attributes of an *omphalus* or world-centre, revered by the whole tribe or nation as its birthplace and the symbol of its identity. In Athelstan's scheme its measure is 6 ft across, which is also the length of the sceptre or rod by which the king metes out or measures out justice. And through the centre of the altar runs a stout, vertical pole, six inches in diameter, that is regarded as an extension of the cosmic axis between the two poles of the universe. This central axis is the only fixed, unmoving component in the world of endless change that revolves around it, and it is therefore the first symbol of divine law that is also constant and unchanging. Its position, erect at the hub of the Saxon royal order, signifies a ritualized society, dedicated to upholding a traditional code of natural law that was revealed by the gods to their ancestors at the beginning of history.

Starting at the centre with the pole, altar and sanctuary, the pattern of the royal court – the mean term in the progression from the human to the divine level – expands in scale to reach its limit at the outer edge of the royal forest-lands, wherein criminals were judged guilty of sacrilege. But that is not the limit of its influence. For, as in all societies where the values and customs of the state capital set the tone for the provinces, the example of the Saxon court was imitated in villages and households throughout the kingdom. This was, in a sense, rule by fashion, but it was a fashion that did not fluctuate, for Athelstan's lawmen were concerned with stability rather than change.

The purpose and effect of their state pattern was to maintain a psychological condition, properly called an *enchantment,* in which ordinary life was exalted – experienced as blessed, divinely ordered and, as far as the pains of existence allow, happy. An enchantment implies chant, and the principal cause of law and order was music and festivals. A round of feast days marked the various stages in the farmer's and hunter's year; and at each of these, at the same spot on the usual day, local people sang the traditional songs and enacted the same rituals as they believed they had always done. Meanwhile, the king with his judges and courtiers progressed around the country, imitating the sun in its annual passage through the zodiac, visiting each of its twelve sections for one month in the year, delivering justice and upholding standards of culture and husbandry. At every spot where the royal court located itself, the sanctuary of the King's Girth was extended into the surrounding countryside.

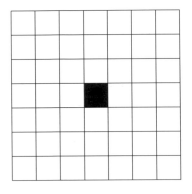

The diagram below that illustrates the King's Girth is not just a clever contrivance of numbers and measures, but gives insight into the science behind ancient statecraft.

1. The area of the King's Girth, whether regarded as a square or a circle, measures 7 miles across. This diagram shows the overall plan. The square is divided into $7 \times 7 = 49$ smaller squares, each of 1 mile. The distance from the side of the shaded squares to the side of the outer square is 3 miles, the 1st measure of the King's Girth.

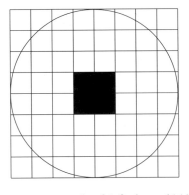

2. This represents the central square, 1 mile wide, in the previous diagram. It is divided into $8 \times 8 = 64$ smaller squares, each of 1 furlong (660 ft). The distance from the side of the central, shaded block of squares to the side of the greater square is 3 furlongs, the 2nd measure of the KG. The 3rd measure, 9 linear acres of 648 ft, lies within and occupies most of the central shaded area. At the very centre is the 24 ft square of the central shrine, the 6ft square of the central altar and the 6 in wide pole at the hub of all.

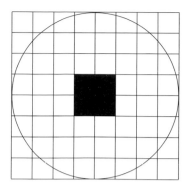

3. So here is the tiny 24 ft-wide square at the centre of the previous diagram, divided into $8 \times 8 = 64$ smaller squares, each of 3ft or 1 yard. It contains, shaded, the 6 ft altar and the central 6 in wide pole.

It was indeed a science, properly so called, based on the demonstrable truths of number and geometry rather than on expediency and opinion. For those who can see Beauty in her abstract, essential form, through the harmonies of number and traditional measure, many delights and wonders can be found among the areas and ratios in the King's Girth diagram. And those who have followed the discoveries of classical Pythagorean science in the designs of megalithic circles can perceive a tradition that has persisted in the British Isles from Stone Age antiquity through Saxon times and into the Middle Ages. This tradition, and even the fact of its existence, is scarcely recognized today. But those who wish can study it, to the benefit not only of themselves but of their entire community and (human) race.

The conclusions that ensue from analysis of the King's Girth are several and radical – not altogether in accord with modern notions and prejudices. The Saxons were not just rough and boorish. Their rulers were learned men in the classical sense; educated in number, measure, geometry, astronomy, land surveying and the classical sciences. Behind that level of education lies another, relating to a form of science more subtle and valuable than any that is known today. The evidence of archaeology and anthropology is unanimous; that the ultimate purpose of this further science was for divination – for communication with local spirits and those of the dead, and for attracting grace, divine blessings or plain good luck to one's self and surroundings.

The King's Girth exemplifies this science. It depicts a little universe, self-contained and with its own laws, customs, music and traditions. Whether it is a kingdom, province, manor or an individual household, its basic pattern is the same, and this pattern is not only expressed in land measures. It lies behind every institution and influences the whole of society, including its religion and world-view. In archaic England, from pagan into Christian times, the centre-point of every judicial or legislative assembly was a tree, rock or upright pole. Around it was an open space, fenced off and dedicated to the supreme deity who upheld the principle of free speech within the sanctum. It was a place of ceremony, legislation and justice. Democracy has its natural roots there, for everyone on the manor attended its court and was party to its legal and administrative proceedings.

An essential feature of this social and constitutional order is its numerical basis. In classical, civilized societies the dominant number in state and religious institutions is Twelve, representing the 12 gods in the zodiac, 12 solar months and many other examples of twelve-fold order. The values symbolized by this number are solar, rational and imperial.

Seven, on the other hand, is the natural number of the soul and spirit, of oracles and inspiration. It is the characteristic symbol of archaic and nomadic societies, where the 7 planetary gods, the 7 stars of the Great Bear and the 7-day week of the moon dominate life and ritual. The relationship between these two number symbols is illustrated in music; seven being the number of notes in the simple music of the shepherd's pipe, whereas in the sophisticated scale of religious chant the number is twelve.

Seven, the number of miles in the overall width of the King's Girth, is indicative of a very ancient measure, not merely of classical origin but also a survival from the old Nordic culture that gave birth to the Celtic, Viking, Saxon and other traditions.

Yet behind this apparent Nordic source is something greater, a unified code of knowledge that can be recognized in the primordial traditions of all nations. This leads to the big question in ancient history. Was there once, many thousands of years ago, a universal culture, embracing all humanity?

Many speculative writers today are thinking in that direction, but the evidence is against it – one objection being that the origins and heydays of every nation and culture are in different periods of history. There seems to have been a succession of cultural revelation – a more interesting conclusion because it implies that revelation is an ever-active principle, as much so today as at any other time. Basically it is a code of law – not a legalistic or moral code nor a set of precepts but a numerical pattern – an abstract expression of the universal culture. It is that constant, divine law which is symbolised by the unmoving axis of the cosmos and the upright pole at the centre of the Saxon manor.

So the quaint old measure of the King's Girth actually introduces us to the educated Saxon mind, to the religiously ordered world-view established in it and to the corresponding order of society. Further pursued towards its roots, it leads us through knowledge of ancient science to confront the most serious question of all. Is this world an artefact? The premise of the question is this. The noble Saxons, in common with the priests, initiates and philosophers of all antiquity, understood the world to be a divine creation, perfectly composed in every detail of number, measure and weight. That is the traditional, ever-recurring, Platonic view of things. It occurs spontaneously to anyone who studies the ancient sciences and the subtle, comprehensive code of number that underlies them all. And it is now occurring in modern physics.

According to the established Anthropic Principle, it so happens that the balance of forces in the universe is, against all conceivable odds, exactly as it would have to be to allow for intelligent life. To many physicists this looks like intelligent design. There is only one self-conscious form of intelligence that we know of – ourselves.

But since we cannot claim the credit for having created the universe, we have to acknowledge another, higher intelligence. That is how we recognize a Creator. And that is why they had a six-inch wide pole at the centre of the Saxon manor.

Appendix VIII

The Mystery of 42

Charles Lutwidge Dodgson lectured in mathematics at Oxford: hence the frequent references in the *Alice* books to measurement – see An Anthology. This curious number 42 also appears in his long nonsense poem *The Hunting of the Snark*: "The helmsman used to stand by with tears in his eyes – he knew it was all wrong, but alas! Rule 42 of the [Naval] Code, 'No one shall speak to the Man at the Helm'" And again: "He had forty-two boxes, all carefully packed..."

You may recall, too, that Douglas Adams (who died in 2001) wrote in *The Hitch Hiker's Guide to the Galaxy*: "The Answer to the Great Question of... Life, the Universe and Everything...[is] Forty-two."

Evidently, the product of 6×7 possesses some compelling, mystical meaning. The Old Testament uses it continually. *Revelation xiii, 5*: 'and power was given unto him to continue forty and two months.' *Judges xii, 6*: 'Then they took him, and slew him at the passages of Jordan; and there fell at that time of the Ephraimites forty and two thousand.' *II Kings ii, 24*: 'And there came forth two she-bears out of the wood, and tare forty and two children of them.' *II Kings x, 14*: 'And they took them alive, and slew them at the pit of the shearing house, even two and forty men; neither left he any of them.'

Yet again, II Chronicles xxii, 2: 'Forty and two years old was Ahaziah when he began to reign...' And Matthew, like Luke, begins the story of Jesus (i, 2-16) with a genealogy tracing his family back to Abraham through 42 generations. *Why forever 42?*

Bear in mind, too, that the circumference of the Earth = $42 \times 6,000 = 252,000$ ancient Greek stadia; the length of 1 meridian degree = $42 \times 5,000 = 210,000$ cubits, and 42 was the qualifying age for the Roman Imperial rank of Consul...but this can become obsessive!

Three trifling instances, however: in Scotland, the home of curling, the traditional length of an 'end' (or a 'rink') was 42 yards and the weight of the stones 42 lb; and in the grocery trade it is common to sell cans, bottles, fruit, etc. in 42s (packing 6x7) – but that exposes the awkwardness of decimal pricing, so the goods are often offered at a price for *40 plus extra 2 free*!

By the way, England possesses 42 Anglican cathedrals. And an Honorary Member, Christopher Martin-Jenkins, points out that the most important Law of Cricket, dealing with fair and unfair play, is Law No. 42.

Finally, to revert to Lewis Carroll, for his amazing story of the gravity train, in chapter 7 of *Sylvie and Bruno Concluded*.

Imagine a railway line running in a straight line from London to Edinburgh or, indeed, from any place on the Earth's surface to any other. Because the Earth's surface is curved, the line cuts into the Earth and so, to an observer at either end, appears to run downhill. But in fact only the first half of the route is downhill. So, if the train is released at one end and runs solely by the force of gravity, it will roll downhill, accelerating until it got half-way when it would start going uphill and consequently slow down. In the absence of resisting forces, it would come to rest at the other terminus. How long does the journey take? Remarkably, this duration is fixed, regardless of the distance between the termini or the weight of the train or the size of the Earth (just as the time of a pendulum's arc is fixed), because the only determinants are *gravity* and the Earth's *density*.

To make the problem mathematically amenable, we assume that the Earth is a uniform non-rotating solid sphere. (Of course, it's not uniform – the heavier bits tend to be in the middle – and it does rotate and it's not quite spherical, but these are marginal considerations.)

Now the equation for the journey time is: *4Gdt2 = 3pi* where 'd' is the Earth's density, 'G' is the universal constant of gravitation and 't' is the required answer. This equation (like every equation in physics) is valid whatever coherent system of units is used. Here, 'coherent' means that if we are using the foot-pound-second system, then speed must be measured in feet per second rather than miles per hour. The value of 'G' is expressed by a formula (too complex to elaborate here) which, applied to the simple term of 347$\frac{1}{4}$ lb/cu.ft to represent the Earth's density, yields a figure of 2,520 for 't'. So that's the journey time in seconds: just 42 minutes – always!

[Most authorities give the Earth's density as 344.4 lb/cu.ft, so the journey time would actually be a few seconds more than 42 minutes, but why spoil it!]

The question is: did Douglas Adams know all this?

Appendix IX

The Standards of Scotland

abstracted from a paper presented by LAWRENCE BURRELL,
DPA, MIWMA, to the Scottish Branch of (what became)
The Institute of Trading Standards on 14 October 1960

Although it is now more than 250 years since the standards were officially abolished, they still present a certain fascination to anyone interested in the history of weights and measures. There is, so far as I am aware, no authentic history available on the subject, though many references are to be found scattered through a number of books published in the 17th, 18th and 19th centuries. This lack of information, coupled with the many apparent and in some cases inexplicable discrepancies or mistakes, negligence or bad workmanship by the assizers and makers – and last but by no means least the innumerable local and customary measures that were in use throughout the country – makes the subject one of personal conjecture. The final standards, approved in 1618 and ratified in 1621, were abolished by the 17th Article of the Treaty of Union in 1707 but, appreciating the 'canny Scot', the antipathy of many to the Treaty – and his very conservative attitude to changes in local customs – it is not surprising that the Article was not strictly enforced [*any more than compulsory metrication is today!*], and the standards remained in fairly general use, especially in agricultural areas, up to the 19th and even into the 20th century.

The position was never really clarified till the promulgation of what in many ways might be termed the basic Act of the present system – i.e. the Act of 1824 – which set up a Jury in each County to establish the relationship between the old and new systems. Even this was not completely successful and it was not till the 1835 Act, Section 6 of which abolished, and made an offence the use of, local and customary measures, that some sort of uniformity appeared out of chaos. Even today we find a similar Section in the principal Act of present-day Weights and Measures. It is of some interest to note that the 17th Article of the Treaty of Union remained in force till it was repealed by the Weights and Measures Act of 1878!

Many examples remain today of customary terms having direct reference to the old standards, the most common being the sale of coal in Dundee by the $\frac{1}{2}$ 'met' of 84lb (6 imperial stones). The 'met' was originally $10\frac{1}{2}$ Scotch Troye or Dutch stones or 168lb (12 stones). The sale of meal or flour by the bag of 140lb is 8 stone Scotch Troye or the old meal 'boll'. Grain is still sold by the 'quarter' of 280lb which is 16 stone Scotch Troye − 20 imperial stones.

The use of the 'porter' gauge in weaving jute is still well known. Jute is woven with various numbers of threads to the porter of 1.85". Cloth of 20 porters is thus 37 inches – the Scottish 'Ell'. [All this only 41 years ago!]

Most of the difficulties attached to the change-over can best be described by a quick glance at the 1826 Jury findings for the County of Forfar. "The pound of Scotch Troyes weight commonly called Dutch weight contains 7,608.9496875 imperial standard grains and the proportion to the imperial Avoirdupois pound is as 1.0869928 to one." A simple conversion!

It might be of some value and interest to examine the words Troy, Troye, Trois and Tron. They appear in innumerable Acts in both countries and whether they are one and the same has a very important bearing on the relationship between the standards of James I and James VI. There is no doubt whatever that Troye and Tron are completely different and were not linked together in any way other than by an arithmetical ratio for purposes of comparison. Chaney in his *Our Weights and Measures* (1897) gives this ratio as $1\frac{1}{6}$. Tron had a fairly wide meaning and, in addition to being a series of weights it was also the public weighing machine. It was necessary to erect weighing machines in various localities for weighing sacks of wool and fleece (on which David II c.1360 imposed a duty of one penny per sack to help pay ransom to the English).

It is something more than coincidence that the word Troy appears in the Scottish, English and French systems. What is its real origin? There is no reference to Troy in Scotland until 1618 – the final standards – all the weights from James I until then being referred to as Trois. It is practically certain, however, that Trois and Tron (unlike Troye and Tron) were the same. As far as can be ascertained, the first reference to Troy in England is an Act of Henry II (450 years earlier): *pur la libre de troy once*. So the most common theory is from the French town of Troyes – an old commercial centre.

Dubious alternatives are from *Trona* meaning 'a beam' or from the ancient monkish name given to London of *Troy Novant* founded on the legend of Brute (King Arthur) – so Troy weight was London weight. Yet another theory was that the origin of weight and money in Scotland as in England, France and in other countries derived from the weight of grain and were measured by the penny ounce, penny shilling and pound; so three in number – hence *trois* – simple!

Here are the probable figures for the original Troye and Tron (Trois) weights, together with official figures for the standards based on the 1826 Jury findings.

TROYE WEIGHT

(a) based on 'Lanark Stone' 121,743.2 grains (b) based on original 122,400 grains

	Grains	Drops	Ounces	Pounds	Stones
a)	29.72	1			
b)	29.9				
a)	475.56	16	1		
b)	478.1				
a)	7,608.95	256	16	1	
b)	7,650				
a)	121,743.2	4,096	256	16	1
b)	122,400				

TRONE / TROIS WEIGHT

(a) based on 'Edinburgh Tron' pound (b) based on original 10,240 grains

	Grains	Drops	Ounces	Pounds	Stones
a)	35.59	1			
b)	40.00				
a)	601.40	16	1		
b)	640.00				
a)	9,622.67	256	16	1	
b)	10,240.00				
a)	153,962.67				1
b)	163,840				

SCOTTISH LIQUID MEASURE

'Stirling pint' = 26,306.982 grains

Weight of water	Imp.Cub.Ins.	Gill	Mutch	Chop	Pint	Gallon*
0.235	6.5	1				
0.94	26.05	4	1			
1.88	52.10	8	2	1		
3.76	104.20	16	4	2	1	
30.07	833.63	126	32	16	8	1

*This differs totally from the imperial gallon (a measure of volume, not of weight) which – 160 fluid ounces and holds 277.12 cubic inches.

SCOTTISH LINEAL MEASURE
Standard 'Ell' = 37.0598"

Imperial Inches	Scotch Links	Feet	Ells	Falls	Chains	Furlongs	Mile
1.002							
8.89	1						
12.02	1.35	1					
37.06	4.17	$^{31}/_{12}$	1				
222.36	25.00	$^{37}/_{2}$	6	1			
889.44	100.00	74	24	4	1		
8,894.35	1,000.00	740	240	40	10	1	
1,976.50 yards	8,000.00	5,920	1,920	320	80	8	1

There were also:

a) the Scottish Dry Measure for Wheat, raised from the Standard Wheat Firlot of 2,214.322 cubic inches, whereby the largest unit was the Chalder, 1 of which = 16 Bolls = 64 Firlots = 256 Pecks = 1,024 Lippies = 1,360 Pints = 141,716.6 imperial cubic inches; and

b) the Dry Measure for Barley and Oats, raised from the Standard Barley Firlot of 3,230.3 cubic inches, whereby 1 Chalder = 16 Bolls = 64 Firlots = 256 Pecks = 1,024 Lippies = 1,984 Pints = 206,739.55 cubic inches (and all without calculators or computers!)

Appendix X

Binary Arithmetic and Measurement

The death in April 2001 of the American mathematician Claude Shannon gained little attention, yet it was he who, as a 24 year-old student in 1940 showed that all information can ultimately be boiled down to just 1s and 0s, which he termed *binary digits* or *bits* – and so the Age of Information Technology was born. His 1948 paper on The Mathematical Theory of Communication has been hailed as the 'Magna Carta of the Information Age'. It laid the foundations for today's digital revolution, from pristine CD recordings to the internet.

Like so many topics in this book, binary arithmetic is a subject about which most readers know something but not as much as they would like – nor, in this particular case, as they ought to know, in view of the importance of binary arithmetic today as the basis of computer mathematics, on which modern technology largely depends. John Strange has contributed the following article, which shows that customary measures are as futuristic as immemorial.

1. Scales of Notation

As human beings, we learn by fitting new ideas and information into patterns formed by ideas and information with which we are already familiar; whereas computers learn only by stringing together logical deductions.

For many hundreds of years the symbol '0' (or some other symbol serving the same purpose) was used as a separator to distinguish, for example, 203 – two bundles of ten bundles of ten plus three units – from 23. The Babylonians introduced a separating symbol in the 2nd century BC and the Hindus did so early in the Christian era. The Hindu system was brought to

Europe by the Arabs, who gave us the word 'cipher', and is the system universally used today. It was again the Hindus, in about the 6th century AD, who first recognized '0' as a number in its own right and not merely as a separating symbol.

The reasons why we use ten as the base for our system of numeration are simply that the Hindus did and that to change now would cause a huge upheaval in return for little benefit. But that is not to say that other systems have not been tried and left their mark. '20' was popular 4,000 years ago and the 'score' is well-rooted in our language. The Babylonians were using a system based on 6 and 10. It became so well entrenched that, in spite of the futile French effort to decimalize time, there are still 60 seconds to the minute, etc. The number 12 has many advocates, principally in the learned Dozenal Society in Britain and the USA.

If 8 children share 42 sweets among themselves, then each one gets 5 sweets and there are 2 left over. In mathematical terminology, when 42 (the dividend) is divided by 8 (the divisor), the quotient is 5 and the remainder is 2. This equation 'dividend = (divisor × quotient) + remainder' always holds good and is fundamental to a sizeable chunk of algebra that includes arithmetic and systems of numeration in particular.

Now if we use 8 as the base of our number system, we could write 52 for the number 42: i.e. 5 lots of 8 + 2.

If there were only 7 children instead of 8, each one would get 6 sweets with none left over. So, in the scale of 7, forty-two is written 60. Again, if there were 6 children, each would get 7 sweets. But we don't write 70 for forty-two in the scale of 6 because, in this scale, 7 is written 11: i.e. one lot of six plus one.

The working can be set out as follows:

6	42		In the first column we write the divisor (in this case 6) in the
0	7	0	second we write the dividend and beneath it the quotient, which
	1	1	becomes the dividend for the next division, and in the third column we write the remainder.

The digits appearing in italics give us our answer. In the scale of 6, forty-two is written 110: i.e. 1 lot of 6×6 plus 1 lot of 6 and no units. A quasi-mechanical procedure for calculation as above is called an algorithm in honour of the 9th century Arabian mathematician Al Khwarismi. Using an algorithm, we can now calculate forty-two in the scales of 5, 4, 3 and 2.

5	42			4	42			3	42	
5	8	2		4	10	2		3	14	0
	1	*3*			2	2		3	4	2
									1	*1*

giving 132   giving 222   giving 1120

2	42	
2	21	0
2	10	1
2	5	0
2	2	1
	1	0

giving 101010

as an exercise, the reader is now invited to write 42 in the scale of 9

A difficulty arises from scales with a base (or 'radix') greater than 10: the lack of symbols to represent these numbers. However, it is easy to see that forty-two is written 39 in the scale of 11 and 36 in the scale of 12.

It is remarkable that while computers, which work exclusively in the binary scale (base 2), are pervading so many aspects of our daily lives, the teaching in our schools is moving away from everything that is not decimal!

The large prime numbers that computers sometimes find are usually expressed using only the digit 1. For example, the following binary numbers are prime: 11, 111, 11111, 1111111; and from 1876 till 1951 the binary number consisting of 127 ones was the largest known prime.

Advocates of the base numbers 12 and 60 point out that 12 is divisible by 2, 3 and 4 while 60 is divisible by 5 as well. But the simplest base is obviously 2. Its main inconvenience, however, lies in the fact that nearly all numbers need about 3.322 times as many digits to be written in base 2 as in base 10. [The mathematically minded reader may feel that this statement lacks precision, so here is a technical exposition. For any positive integer 'n', let $d(n)$ be the number of digits needed to write it in the denary (or decimal) scale and let $D(n)$ be the number of digits needed to write it in the binary scale. Then the limit of $D(n)/d(n)$ as 'n' tends towards infinity is $\log 10/\log 2$.]

Returning to the representation of 42 in the ternary scale (base 3), let us do the following sums: $1120+10 = 1200$; $1200+100 = 2000$; $2000+1000 = 10000$. Each time that a 2 occurs, we have added 1 in the same column to give 3, which in the ternary scale is written 10. The total effect of what we have done is to add 1110 to 1120, giving the answer 10000. Or, $1120 = 10000 - 1110$. It is clear that the method is general and so we have the following theorem: "Any number can be written as the difference between two ternary numbers, neither of which uses the digit 2."

Finally, for those that like that sort of thing, here's an original puzzle. The following addition is done in the scale of 7 and each of the letters occurring in it represents a different digit from 0 to 6. Find the value of each letter!

$$
\begin{array}{r}
\text{THIS} \\
+ \text{IS} \\
\hline
\text{EASY}
\end{array}
$$

2. ILLUSTRATIONS

Division by 2 is quite an easy matter in practice – we can all double a length of string or fold a sheet of paper. Division by 4 is no harder, for we merely divide again by 2. Reference to a good dictionary shows how extensively the word 'quarter' occurs in our language. But division by other numbers such as 3 or 5 is a different matter altogether. Construction of a 'compass rose' provides a good example of repeated division by 2. From the quarters marked N, S, W and E – the cardinal points – we proceed to eighths by the addition of NE, SE, SW and NW; then to sixteenths by the further addition of NNE, ENE, ESE, SSE, SSW, WSW, WNW and NNW.

Finally, to divide the circle into 32 parts, 8 more diameters are drawn to produce the 'by' points, in naming which the golden rule is 'Never start with a 3-letter point but always end with a cardinal. Thus, the 9 point for the quadrant from N to E are: N, N by E, NNE, NE by N, NE, NE by E, ENE, E by N, and E.

In the old days, each point was divided into quarters, in naming which the same rule applies – never start with a 3-letter point but always end with a cardinal. Here are the first few, working from North towards East. N, N $^1/_4$ E, N $^1/_2$ E, N $^3/_4$ E, N by E, N by E $^1/_4$ E, N by E $^1/_2$ E, N by E $^3/_4$ E, NNE, NE by N $^3/_4$ N, NE by N $^1/_2$ N, etc. This way of naming the points of the compass makes the precise direction indicated immediately clear, whereas nowadays a seaman in the bowels of a warship who has been told to steer 'two three zero' has to think for a moment or two before recognizing that he was steering a course that's roughly south-west!

While on nautical matters, it is worth mentioning that all British ships are divided into 64 'shares'. In the binary scale, 64 is written 1000000!

A further example of repeated division by 2 is provided by the modern system of paper-sizes, starting with a sheet of 46.8192in × 33.1062 = 1,550 sq.in, or 1.18921m × 0.840896m = 1sq.m, then halving to A3 size and halving again to A4 (11.7048in × 8.27654in) and so on, but always retaining the same proportions.

As a third illustration, suppose that we have an ordinary set of scales in which the weights are placed in one pan and the object to be weighed in the other. Let us suppose that we have a weight of one unit: an ounce, a pound or anything. We can now deal with objects of weight 0 and objects of weight 1: 0+1 = 1. If a weight of 2 units is added, we can double our range: 0+2 = 2 and 1+2 = 3. We can now add a weight of 4 units and make another table:

$$0 + 4 = 4$$
$$1 + 4 = 5$$
$$2 + 4 = 6$$
$$3 + 4 = 7$$

It is clear what's going to happen next. We introduce a weight of 8 units, allowing us to weigh up to 15 units, and then add a weight of 16 units, and so on = each new unit being twice the last one.

Now 16 drams = 1 ounce and 16 ounces = 1 pound, or, in the binary scale, 10000 drams = 1 ounce and 10000 ounces = 1 pound. Let us therefore suppose that the dram is our initial unit. We saw earlier that forty-two is written 101010 in the binary scale. Now 101010 = 100000 + 1000 + 10, and if we convert these numbers back to the denary scale we get 42 = 32 + 8 + 2.

Therefore, to measure a weight of 42 drams we use the 32 dram (2 ounce) weight, the 8 dram ($^1/_2$ ounce) weight and the 2 dram weight. In such a context, the binary scale is actually simpler and more natural – though less familiar – than the denary scale. So what weights would we use for a thousand drams? The answer is provided by writing 1000 in the binary scale.

2	1000	
2	500	0
2	250	0
2	125	0
2	62	1
2	31	0
2	15	1
2	7	1
2	3	1
	1	1

The algorithm shows that 1,000 in the binary scale is 11,1110,1000. The number has been divided into blocks of four digits to correspond to pounds, ounces and drams.

Reverting to decimal numbers, the binary number shows that we need the following weights to weigh 1,000 drams: 2lb, 1lb, 8oz, 4oz, 2oz and $\frac{1}{2}$ oz.

If you bought a set of metric weights for your kitchen scales, you would probably get 9 weights: a 5 gram, a 10 gram, 2 × 20g, a 50g, a 100g, 2 × 200g and a 500g. The repetition of two of the weights clearly indicates a deficiency. With these weights you can weigh up to 1.105kg (nearly 2lb 7oz) at intervals of 5g (about 3 drams): in all, 222 different weights – including 0.

In contrast, it is easy to see what we could do with 9 British weights, for – as we discovered earlier – the introduction of each new weight doubles our range. The progression is thus: 1,2,4,8,16,32,64,128,256,512. Therefore, with 9 British weights we could weigh up to 511 units at intervals of 1 unit: in all, 512 different weights – including 0.

If, for example, we chose the 9 weights from 2 drams to 2 pounds, we could weigh up to 3lb 15oz 14 drams, at 2 dram intervals. That is just $\frac{3}{5}$ths more than by metric weights, and at closer intervals: i.e. both the range and precision are greater with imperial units. In binary notation (and using the binary point), that maximum weight is equivalent to 11.1111111 lb, these 9 'ones' showing that all nine weights had been used.

All we have done here illustrates what the Commissioners said in their first report of 1819 (Appendix III): "...the continual division by two renders it practicable to make up any given quantity, with the smallest possible number of standard Weights and Measures, and is far preferable, in this respect, to any decimal scale."

By way of a footnote, let us suppose that some joker has provided us with weights of 1 unit, 3 units, 9, 27, 81 units, etc.

– each unit being 3 times the one before. This is an entirely fantastic idea, of course, but interesting nevertheless, if only to illustrate the theorem about *ternary* numbers at the end of the first half of this article. We can still manage if we are allowed to put weights into both scale pans. For example, as we have seen, 42 in the ternary scale is 1120 and this may also be written as 10000 − 1110.

In other words, reverting to decimal notation, an object of weight 42 units can be weighed by placing the 81 unit weight in one pan and, in the other pan, the 27, 9 and 3 unit weights alongside the object to be weighed.

For an object weighing 2000 units, we first express 2000 in the ternary scale:

3	2000	
3	666	2
3	222	0
3	74	0
3	24	2
3	8	0
	2	2

So 2000 in the ternary scale is 2202002. Proceeding as we did at the end of the first half of this article: 2202002 + 101001 = 10010010. Therefore: 2202002 = 10010010 − 101001. In other words, reverting to the decimal scale, an object of weight 2000 is weighed by putting weights of 2187, 81 and 3 units in one pan and, in the other pan, the object to be weighed together with weights of 243, 27 and 1 unit.

Appendix XI

A Humorous Adversary

Unlike metricksters at home, their American counterparts do possess a sense of humour. So, for light relief, here is a comment from Mr R W Bemer of Honeywell Information Systems in Phoenix, Arizona, writing in the journal, *Mathematics Teacher*: "If the average horse is 15.84 hands tall, and we know that one mile equals 15,840 hands, is one mile equal to 1,000 horses? And that's just for the statute mile: how about sea-horses and the nautical mile?

And everyone knows that a barrel contains 31.5 gallons, except if it contains petroleum products, in which case it's 42 gallons; and except for malt beverages, in which case it's 31 gallons. Of course, these barrels are for liquids, but the barrel is also defined for dry content, measured in quarts rather than gallons, and everyone knows that a barrel for dry content contains 105 quarts, except if it contains cranberries, in which case it's 87 quarts. And, in seeking the lowest common denominator for linear measure, the statute mile can be broken down to 63,360 inches and even to 84,480 digits and finally to 4,055,040 ounces. You think ounces per mile is a mistake? Not so: the ounce is equivalent to $^{1}/_{64}{}^{th}$ of an inch for measuring the thickness of leather."

What laughs! But why shouldn't the thickness of leather be measured in 64ths of an inch and $^{1}/_{64}{}^{th}$ called an 'ounce'? There's an excellent reason for it in the leather trade, and it doesn't conflict with anything else. And who doesn't know the difference between a four-legged horse and a sea-horse, or is ever likely to confuse a barrel of oil with a barrel full of cranberries?

Besides, the respect earned by his humour was lost when he went on to complain that "Fractions are difficult for the French.

The only fractions the French use is one half – they will order a demi-litre of wine, but a smaller amount is not a quarter litre (2.5 decilitres) but an even 2.0 decilitres. [Not so: *un quart de vin* is common (25cl) and so are quarter bottles of 18.75cl.] Our (US and UK) fractional system is the awkward octal or hexadecimal one of computers; whereas theirs goes 2, 1, 0.5, 0.2, 0.1, 0.05, and so on, to form a sequence that repeats in tens." Such an erratic system of numbering would defeat any computer! So what's at fault is computer arithmetic, not the metric system – and ignorance of fractions (together with its inevitable consequence, which is an incapacity for mental arithmetic) is a sign of superior intelligence!

Anybody can so easily mock the names of imperial units, and point to apparent anomalies, precisely because they all directly relate to the human form and the practical uses for which they are designed, all of which have their fascinating vagaries and idiosyncrasies, whereas metric units relate to nothing whatever in the real world – they are abstract, without meaning or feeling, devoid of cultural or historical interest, and therefore immune to wit and joy.

Thomas McGreevey in his book *The Basis of Measurement* (Picton Publishing Ltd, Chippenham, 1997: ISBN 0 948251 840) recalls the story of the scientist who informed the US Armed Forces that the estimated velocity of a new type of rocket was $3 \times (10$ to the power $5)$ centimetres per second and then, in response to their request for a conversion into 'English' units, reluctantly advised: "$1.8 \times (10$ to the power $7)$ furlongs per fortnight." Good fun!

Appendix XII

Questions and Answers

A. Weighing the Earth

A Mr Chambers replied in *The Times* (27 May 2002) to the question: "Can planets be weighed – if so, how much does the Earth weigh?

Weight is the measure of the force by which mass is pulled towards the centre of the Earth. Arguably... the *'weight'* of the Earth is zero! In Peru in 1740, Pierre Bouger (1698-1758), together with Charles Marie de la Condamine (1701-74), made measurements which he published in *La Figure de la Terre* (1749). A more accurate result was obtained, using the same principles, by the Astronomer Royal Nevil Maskelyne (1732-1811) in 1774. He chose the Scottish mountain Schiehallion as relatively isolated and regular in shape, and compared the difference between a 'plumb-bob' vertical and the astronomical zenith on the north and south sides of the mountain. In both cases the plumb-bob was deflected slightly toward the mountain by the attraction of its mass. He concluded that the density of the Earth was about 4.5 times that of water. This figure is now known to be 5.517, which makes the mass of the Earth about 5,980 million million million tonnes. The times and distances involved in the orbits of the planets, together with the laws of physics, make it possible to calculate the masses of other planets relative to the Earth. Jupiter has 317.89 times the Earth's mass, whereas Pluto has only 0.002 times..."

But a Mr Milner (4 June) objected: "Nevil Maskelyne's Schiehallion experiment...and those of Bouguer and de la Condamine, were not intended to measure the mass of the Earth but to find the universal gravitational constant (Newton's 'G'). To do this, Maskelyne assumed the density of the rock...in order to estimate its total mass. Had he been using his experiment to find the mass of Schiehallion, he would have needed to know 'G' already. The true value of 'G' was later found...by Cavendish using a torsion balance."

Furthermore, from a Ms Penney. "The first attempt to weigh the Earth took place in Birmingham in 1890, when John Henry Poynting, first Professor of Physics at Sir Josiah Mason's Science College (forerunner of the University of Birmingham) carried out an experiment which he called 'measuring the gravitational constant'. He used a balance made by the German firm Oertling in 1880. This is now in the National Physical Laboratory. The experiment took 12 months to complete and was performed in the college basement after the floor had been strengthened. Poynting's room was above and for observation purposes a large hole had to be cut in the floor. Until the building was demolished the site was marked with a plaque, saying: 'Here the Earth was first weighed.' Poynting's calculation worked out at 12,600,000 million million million *lbs*..."

B. DAYS IN MONTHS

The Times on 21 May 2002 had carried this further response (correcting an earlier one), from a Mr Cleverley, to the question: "Why was February given 28 days (apart from leap year, of course)? Wouldn't it have been easier to give January, February and March 30 days each?"

"No civilization, ancient or modern, ever had a ten-month year. December was the tenth month because the year started in March (as it did in England until the 18th century – see Time). February was, therefore, the last month of the year. Pre-Caesar, March, May, July (then called Quintilis) and October had 31 days; the rest had 29 except the last, February, which had 28. Since this added up to only 355 days, an extra 22 or 23 days were added to every other year. This was done by including an extra month (Intercalaris) starting on what would have been the

24th of February and having 27 or 28 days.

A 4-year cycle therefore consisted of 355+378+355+377 – still leaving four days too many. The resultant discrepancies were taken care of by *ad hoc* adjustments from time to time....Caesar abandoned this system, substituting the 365-day year with one leap day at the end of the year every four years. The ten extra days were distributed over the existing months, giving 31 to January, March, May, July, October and December, and the rest 30, except for February, which still had 29, with 30 in leap years Later, when Augustus had the sixth month (Sextilis) named after him, it was felt it should have another day to equal July: the extra day was taken from the end of the year – the last day of February."

This elicited yet a further response (23 May) from a Mr Oliver. "The (very) early Roman calendar did have just ten months, with anuncounted period covering what is now January and February. If February had always been there as the twelfth month, the Romans would logically have called it Duodecember. Mr Cleverley is right in that this gap was subsequently filled with January and February, and a leap month every second year after February 23, but it seems that this actually resulted in an average year of less than 365 days, because the Romans do not seem to have gone back to the last days of February after the leap month but went straight to the first day of March.

And to be very picky: it was September and November, rather than October and December, that had 31 days in the early Julian calendar – these lengths were only reversed as part of the changes when July and August were thus named."

C. Sunrise and Sunset

The Times (4 November 2002) answered the question: "When we in the UK see the Sun setting, how far away are the people who, at that moment, can see it rising?"

The answer depends on the latitude of the observers, their height above the visible horizon and the time of year. The farther north you are in the northern summer, the closer together the observers of sunset and sunrise will be. The maximum distance the people can be apart occurs on the equator at the equinox. The longitude distance of the two places is slightly greater than 180 degrees owing to the diameter of the Sun and the bending of light in the Earth's atmosphere, which allows both observers to see the Sun when it is below the horizon. Measured along the sunlit part of the equator, these places are separated by 181deg.40min of longitude. Their distance apart is less along the unlit part: 178deg.20min, corresponding to about 12,300 statute miles.

For places on the same latitude, the longitude difference can be estimated from the published times of sunrise and sunset. At latitude 51deg. north on [31 October 2002], sunset was 9hr.53min after sunrise. This corresponds to sunrise taking place 148deg.15min westwards. So while you are watching the Sun set from Westward Ho!, sunrise is being seen by a fisherman in the Bay of Alaska near Kodiak Island. In midwinter, sunrise would be seen from Vancouver Island and in midsummer over the Sea of Baykal near Irkutsk. [*Kodiak Island is further north than Devon; but these newspaper articles should not be taken too seriously.*]

D. Heat from the Sun

The Sunday Telegraph on 19 July 2002 answered the question: "The Earth is farthest from the Sun in July, so why is this the hottest month of the year?"

"It's certainly true that the Earth is farthest from the Sun about now – about 3 million miles farther away than we are in January – and this does mean that we get about 6 percent less heat from the Sun now than in winter. [But] the real explanation lies in the fact that our planet is tilted towards its orbit at about 23 degrees. Around June and July, the northern hemisphere is tilted towards the Sun, so its rays strike our part of the world at a relatively steep angle, making their heating effect per square foot relatively high. During our winter, however, our hemisphere is tilted away from the Sun, and its rays strike the ground more obliquely, thus diluting the amount of heat that hits each square foot. Down Under, of course, it's the other way around. By that reckoning, our hottest day of the year should be at the summer solstice, about 23 June. In practice, it takes a while for the sea currents that flow past our country to catch up with events overhead, and so it's usually July before we feel the benefit (and often not even then)." [*BWMA doubts the meaning of this assertion that our planet is tilted towards its orbit!*]

E. RAINFALL AND AREA

On 22 September 2002, *The Sunday Telegraph* answered the question: "What is the relationship between the amount of rainfall expressed in inches and the actual volume that falls per unit area?"

The volume of rain depends on the area over which it falls, and it can be a truly astonishing amount. An inch of rain, over an area of only one square mile, amounts to about 15,000,000 gallons or 66,000 tons of water. That is equivalent to about 260 gallons unloaded onto a tiny garden of 500sq.ft. The rate at which nature can deliver these enormous volumes of water almost defies belief. In the notorious Lynmouth Flood of 50 years ago, Exmoor received almost 9 inches of rain in 24 hours – the equivalent of 6,000,000 gallons of water *per hour* on each square mile.

F. Population and Area

On 20 October 2002, *The Sunday Telegraph* answered the question: "How much space would the world's population take up if everyone stood together?"

According to United Nations estimates, the population of the world is currently around 6,200 million. For the sake of argument, let us give each person sufficient room to 'swing a cat' – say a circular space about 11ft across [roughly 95sq.ft]. Multiplying up, the world's population could comfortably fit into a space of around 21,000 square miles. Sri Lanka, Tasmania or Haiti would all fill the bill. If one insists on cramming us all together to the point of standing room only [say 4sq.ft], then it's possible to pack everyone into a space of just 800 square miles, which is an area the size of the Lake District.

G. The Big Bang

A Mr Slyfield replied in *The Times* on 27 January 2003 to the question: 'We had the Big Bang. The Universe has since been expanding, presumably, in all directions. Does anyone know where the centre is? Where did the Big Bang take place?'

"Like evolution, Big Bang theory is one of the most frequently uttered but widely misunderstood concepts in popular science. The Big Bang was the origin of all space and time – 'space' in this context being the entire Universe – all the energy, matter, fundamental forces and the apparent physical space that separates them. Its first manifestation was an 'initial singularity', a sub-atomic energy state that held the entire universe compressed into a point of infinite density and temperature, yet a point of zero size and volume – the difference between existence and non-existence being, at some stage, practically negligible. The Big Bang didn't actually take place anywhere else but right here; as by definition there was nowhere else for it to happen. Similarly, we are all still in the exact centre but then so is everything else. It is the very fabric of space/time that has stretched 'internally' from its original state but has not actually gone anywhere at all."

H. MOUNTAINS

A Mr McKelvie in *The Times* of 9 January 2003 answered the question: 'What size do hills have to be to become mountains, or does something else define the difference?'

"In Scotland, especially the Highlands, 'the hill' may refer to anything from relatively low grouse moors and sheep grazings to high deer forests of 3,000ft and more. 'Mountain' is a word seldom used by Scottish farmers or sportsmen, but is a ramblers' and climbers' term. [A 'Munro' is a peak of 3,000ft or more above sea level.] In Ireland, however, a 'mountain' may mean virtually any area of heather, even if it is at or below sea level and quite flat! Its designation is similar to that of a 'moss' in the Scottish lowlands."

Then a Mr Bourne quoted from Gilbert White's 1789 classic *The Natural History of Selborne:* "The prospect is bounded on the South-East and East by the vast range of mountains called the Sussex Downs. They include Mount Caburn and Mount Harry, on either side of Lewes, rising to about 600ft!"

Also, a Mr Horton pointed out that: "to the North of Fort William in the Highlands there is a range of mountains known as the Aonachs. There are two main mountains: Aonach Beag (translation – 'little hill') and Aonach Mor ('big hill'), with respective heights of 4,049ft and 4,006ft, the higher being the little hill and the lower the big one – because Aonach Mor is the more massive."

Finally, a Mr Upchurch quoted from the 1984 edition of the *Longman Family Dictionary*: "A hill is a rounded natural rise of land lower than a mountain" and "A mountain is a landmass that projects conspicuously above its surroundings and is higher than a hill". He expressed the hope that this would clear up any confusion.

J. More Mountains

The Sunday Telegraph (18 May 2003) answered the question: "I have heard it said that, contrary to popular opinion, Mount Everest is not the highest point on Earth. How so?"

"It all depends what you mean by 'highest'. The usual definition is height above sea level, in which case there is no doubt that Mount Everest is, at 29,029ft, the highest point on the planet. Some have, however, argued that the highest point should be considered to be the furthest point from the centre of the Earth. In that case, one must take into account the fact that the Earth's rotation causes a bulge at the Equator of about 14 miles relative to the poles.

As it happens, Ecuador's Mount Chimborazo lies within one degree of the Equator, and though its summit is only 20,700ft above sea level, it benefits mightily from the bulge effect compared with Mount Everest, its summit standing more than 7,000ft higher from the centre of the Earth. But Everest still remains unchallenged as the best vantage point on the planet.

The rule of thumb for calculating the distance to the horizon in miles is to take the altitude in feet, add on 50%, then take the square root. With Everest standing only 13,000ft above the immediately surrounding land, that could affect the view from the top. However, the land to the SE falls away fast enough to guarantee an uninterrupted view of about 210 miles from the peak. The effects of atmospheric refraction may even increase this to 230 miles or more."

K. FLYING BULLETS

The Sunday Telegraph (19 July 2002) answered the question: "How hazardous are bullets that have been fired into the air?"

"A bullet from a Kalashnikov rifle weighs only about ¼ oz, but it leaves the gun at more than 1,500mph. This gives the tiny bullet the same amount of energy as a brick dropped from the top of St. Paul's Cathedral. Air resistance does slow down bullets, however, and a bullet fired into the air loses about 90% of its muzzle velocity, returning to Earth with its energy reduced to the equivalent of a brick dropped from a height of 4ft or so."

L. 360° CIRCLES

The same issue of *The Sunday Telegraph* also answered the question: "Why are circles divided into 360 degrees?"

"Even in these days of enforced metrication, time and angles continue to be reckoned in the same units used by the Sumerians more than 4,000 years ago... the Sumerians used a base-60 system of arithmetic. [60 is conveniently divisible by every number from 1 to 6.] Their mathematicians must have discovered that a circle can be divided into 6 equal parts by taking the pair of compasses and marking off points along the circumference equal to its radius. So 6 × 60 = 360... Another suggestion... links it to the calendrical year, Sumerian mathematicians having noted that the number of days in a year is close to 360 – the awkward 5 days left over being declared public holidays!"

M. ORDNANCE SURVEY

We are frequently asked: "Why the *Ordnance* Survey?"

"Nothing to do with 'Ordinance', of course! The Ordnance Survey was set up in 1791 when Napoleon was threatening to invade, and the Board of Ordnance (the Defence Ministry of the day) realized that it had little idea of the geography of Britain's vulnerable South coast. It also needed accurate plans of all the roads along which artillery could easily be hauled. Ten years later, the first inch-to-the-mile map of Kent was published, showing all the vantage points from which to fire artillery, and the escarpments that could be defended.

Within twenty years, a third of England and Wales had been mapped on the 1in. scale. Soon a 6in. to the mile map was needed to calculate tithes for church revenues and also to help work out where to build railways and canals across the country. So the Ordnance Survey and its army of surveyors ensured that Britain became the best-mapped nation in the world. Furthermore, these advanced cartographic techniques were transported to cover most of the Empire, where they were invaluable to the administration as well as to traders and the military."

N. EARTH'S INTERIOR

The Sunday Telegraph on 16 March 2003 answered the question: "If the Earth's interior is molten lava, why does the heat escape up to the surface only at places such as volcanoes – and can we tap into this heat as a source of free energy?"

At its formation around 4,500 million years ago, the Earth was molten, and has since cooled to form a 20-mile crust of more or less solid rock. While the crust is thin…it takes a lot of heat to burn through it. Molten material thus usually reaches the surface through cracks in or between the techtonic plates that make up the crust. The most spectacular demonstration of this is the Ring of Fire, the volcanoes around the Pacific plate. Even so, molten material from the Earth's interior does occasionally burn right through the crust, as the dinosaurs may have discovered to their cost 65 million years ago. This requires the creation of a so-called mantle plume, an enormous finger-like blob of extremely hot magma formed at the boundary between the mantle and the outer core, around 1,800 miles beneath the surface.

Once it has broken away from the boundary, the plume rises to the surface, triggering a colossal outburst of volcanic activity. A mantle plume was responsible for dumping about 250,000 cubic miles of lava into what is now the Deccan Traps of India about 65 million years ago – an event of unimaginable violence, the climatic effects of which helped tip the dinosaurs into extinction. Continental drift has since moved the point at which the plume now reaches the surface to the island of Reunion, in the Indian Ocean.

Attempts have been made to exploit the source of free energy available in the heat escaping from the Earth, equivalent to billions of watts. These 'geothermal' energy schemes usually involve injecting water into pipes set in hot rocks, and using the resulting steam to drive turbines. There is probably more scope for such exploitation – if only governments could be persuaded to take them seriously.

The purpose of this Appendix XII was not only to provide a great deal of interesting and valuable information but also to show how precisely, naturally and easily, our customary weights and measures answer even the most difficult scientific questions. Indeed, they simplify the process, because the units are part of the language – they fit comfortably into the common speech of the answers – whereas the metric equivalents are obviously technical terms, therefore obtrusive and distracting.

Appendix XIII

Wire Gauges

Clearly, if we are dealing with a wire nearly half an inch thick, we don't need the next size down to be only one thousandth of an inch smaller; but when dealing with wires only a few thousandths of an inch thick, then that is the sort of precision we do need. In the American text-book *Conductors for Electrical Distribution*, the author F A C Perrine achieves this by choosing each size to be just over eight ninths of the previous one, so that the sixth size down from any number has reduced by half.

This is the same progression that occurs in the well-tempered musical scale [see Music], since six full tones comprise an octave and the note an octave lower than any given note vibrates exactly half as fast. The British Standard Gauge, however, appears to have evolved purely to serve practical purposes, with no theoretical basis. The following table is taken from *The Boy Electrician* by Alfred P Morgan and Oswald Carpenter (George G Harrap, 1920).

Gauge	Diameter (inches)	Gauge	Diameter (inches)
000000	0.464	20	.036
00000	.432	21	.032
0000	.400	22	.028
000	.372	23	.024
00	.348	24	.022
0	.324	25	.020
1	.300	26	.018
2	.276	27	.0164
3	.252	28	.0148
4	.232	29	.0136
5	.212	30	.0124
6	.192	31	.0116
7	.176	32	.0108
8	.160	33	.0100
9	.144	34	.0092
10	.128	35	.0084
11	.116	36	.0076
12	.104	37	.0068
13	.092	38	.0060
14	.080	39	.0052
15	.072	40	.0048
16	.064	41	.0044
17	.056	42	.0040
18	.048	43	.0036
19	.040	44	.0032

APPENDIX XIV

abridged from Andro Linklater's introduction to
MEASURING AMERICA

see also Surveyors' Measures

East Liverpool, Ohio, sits on the banks of the Ohio River just outside the Pennsylvania border...On the road above the Bell Company's dock, Pennsylvania Route 68 invisibly changes to Ohio Route 38, and trees half hide some signs by the roadside. The place could hardly be more anonymous...The language of the signs is equally undemonstrative. A stone marker carries a plaque headed 'The Point of Beginning' that reads: '1112 feet south of this spot was the point of beginning for surveying the public lands of the United States. There on September 30, 1785, Thomas Hutchins, first Geographer of the United States, began the Geographer's Line of the Seven Ranges.'

There is nothing to suggest that it was here that the United States began to take physical shape, nothing to indicate that from here a grid was laid out across the land that would stretch west to the Pacific Ocean and north to Canada and south to the Mexican border, and would cover more than three million square miles, and would create a structure of land-ownership unique in history, and would provide the invisible web that supported the legend of the frontier with its covered wagons and cowboys, its farmers and gold-miners, and would insidiously permeate its formation into the unconscious mind of every American who ever owned a square yard of soil.

"For the distance of 46 chains and 86 links West", Hutchins wrote in his first description of the territory, "the land is remarkably rich with a deep, black mould, free from stone." He was Robinson Crusoe, landed in an uncharted wilderness, and his purpose was to measure it so that it could be sold.

It is easy to miss the significance of what he proposed to do. Surveying is old – its 5,000-year history originates in riverside communities in the Middle East and in Egypt who needed to mark the boundaries of areas to be irrigated by the annual flooding of the Tigris and the Nile. Measurement is older still, and the process is so universal as not to merit a second thought. Yet without it, the exchange of goods and services cannot take place. Consequently, it is almost as necessary to human society as language, and it occurs in the earliest civilizations, long before the development of writing.

Weights and measures are a 'given'. A pound or a gallon, like a mile or and acre, will be the same from Florida to Alaska. And so will a bushel of wheat and a cord of wood and a hundred other units of measurement. It is a language that is picked up automatically and spoken without conscious thought. Only when it changes, when half a gallon of cola became 2 liters in the 1990s, for instance, or a fifth of whiskey reappeared as 750 milliliters [*it is actually 757.0823568ml*] is there a reminder that nothing is certain about these units after all. They are not a 'given' but an extraordinary construct, and one of the identifying marks of social life. Without the conscious decision to agree on a way of measuring, cooperative activity could hardly take place.

Thus, by measuring out the wilderness, Hutchins would make it possible for someone to buy and own it. This was a revolutionary concept. For centuries the land had been lived in by the Delaware and passed through by the Miami and occupied by the Iroquois, but no one had ever owned it. Indeed, throughout most of the world and most of history, such a possibility was inconceivable. Individuals could certainly own the use of land, whole dynasties might grow up around a particular estate, and the right of each generation to occupy and exploit that parcel of land would be unquestioned, but it could not be owned as a house or a pig was owned. The span of one life was too brief to possess the earth that continued forever.

Territory defined the community, or society, or even the nation that lived there. A monarch or ruler who embodied the nation might own it, but no one else. The idea that land might be treated as property belonging to an individual, to be traded, borrowed against, and speculated in like any other commodity, required a fundamental shift in thinking. The concept had been incubated in Tudor England, and taken shape in colonial America, but only with American independence –achieved just two years before Hutchins's arrival on the Ohio – was it fully realized. This was the magic that Hutchins would introduce into western lands – the transformation of the wilderness into property.

The wand that made it possible was there in the first sentence of his report. The land would be measured in chains and links. In most circumstances, a 'chain' imprisons; here it released. What it released from the billowing, uncharted land was a single element – a distance of 22 yards. Invented by the 17th century English mathematician Edmund Gunter [*but all he actually 'invented' was the division of the ancient chain into 100 links*], it was the surveyor's one indispensable tool, and the fact that its 22 yards were to become integral not only to the game of cricket in his own country but to the town planning of almost every major city in the United States (the lengths of most city blocks are multiples of it) was a tribute to the instrument's usefulness.

Repeated often enough – added, squared, and multiplied – measurement with Gunter's chain would give to the land beyond the Ohio a numerical value that someone could compute in money. But there was a double significance to Hutchins's work. [His] survey began at a critical moment in the history of ideas, when for the first time in 10,000 years those traditional measurements were challenged by systems derived from scientific discoveries about gravity and the size of the Earth [*but these traditional units themselves relate to the size of the Earth!*].

In France and Britain, scientists were devising new decimal units, from which the metric system would shortly emerge, but they lagged behind the United States. Before Hutchins crossed the Ohio, Thomas Jefferson had already invented a decimal system based on the length of the equator, which he insisted should be used in the survey. Nor was he alone in arguing for new, scientific, decimal measures. Of the founding fathers, George Washington, James Madison, James Monroe, and even Alexander Hamilton supported him [*yet they rejected the metric system!*], and all understood that the system used in the wilderness would eventually become the system used by the entire United States.

Thus, what began at the spot [in East Liverpool] was not just a survey but the realization of two of the most potent ideas that have shaped America. The first of these was a unique system of measurement that harks back to the dawn of history and, through the grid that the survey laid across America, was to leave its mark on almost every acre of real estate, every farm and every city block west of the Appalachians. The second, and perhaps more important idea was that any individual could own outright the land that Hutchins and his followers were to measure. Around that revolutionary perception grew a society whose economic system and democratic outlook were unlike any that had come before.

In his poem *The Gift Outright*, Robert Frost grasped at another thought, more powerful still. In the end, owning a parcel of earth can never be quite like owning other forms of property. The land has its own magic, and those who seek to possess it are liable to end up being possessed by it. It was the desire to own this particular land, Frost mused, that made its owners American. These were the great forces that Geographer to the United States Captain Thomas Hutchins set in motion when he first unrolled the loops of his chain at the Point of Beginning.

Appendix XV

The Beaufort Scale

see also Nautical Measures

Between 1805 and 1808 (authorities differ as to the date) Captain – later Rear Admiral – Sir Francis Beaufort devised a scale for describing the force of the wind at sea. The criteria he used were the appearance of the sea and the amount of sail that one of His Majesty's frigates would be carrying. This scale started with number 0 and rose to 12 at which point he stopped because by then his frigate would be under bare poles. Wind speed is still measured 33ft above sea level where the force is fullest.

The equation $V = (1^{3}/_{8}) \times B^{1.5}$, where V is the wind speed in knots and B is the Beaufort number, agrees well with the scale that Beaufort elaborated empirically. The only notable discrepancy occurs between forces 4 and 5. Indeed, if we replace B by $4^{1}/_{2}$ in the equation, we find that V is 15.512…so that force 4 might have been expected to stop at 15 knots and 5 to start at 16. Use of the equation allows us to extend the table as far as we like; it was extended to 17 in the early 1950s.

These descriptive terms are taken from *The Admiralty Manual of Seamanship* of 31 December 1951, but different authorities do vary.

The force of the wind acting on a circular disc of area 1 square foot facing the wind can be expressed in lbs by the formula: $0.0105 \times B3$. Thus, a wind of force 3 on the Beaufort scale exerts a force of only 0.28lbs or $4^{1}/_{2}$ozs on the disc, whereas a whole gale (force 10) exerts a force of $10^{1}/_{2}$lbs. The highest wind speed ever recorded near the Earth's surface was 200 knots, which would be 25 on the Beaufort scale – exerting a force on our circular disc of no less than 159lbs.

Wind force (Beaufort)	Speed in knots	Descriptive terms
0	less than 1	Calm
1	1 – 3	Light airs
2	4 – 6	Light breeze
3	7 – 10	Gentle breeze
4	11 – 16	Moderate breeze
5	17 – 21	Fresh breeze
6	22 – 27	Strong breeze
7	28 – 33	Moderate gale
8	34 – 40	Fresh gale
9	41 – 47	Strong gale
10	48 – 55	Whole gale
11	56 – 63	Storm
12	64 – 71	Hurricane
13	72 – 80	do.
14	81 – 89	do.
15	90 – 99	do.
16	100 – 108	do.
17	109 – 118	do.

Appendix XVI

Wine Levels

To conclude on an appropriately convivial, not to say bibulous, note: the level of wine in the bottle can critically affect the quality of the contents. The various levels are strictly identified and defined. For all wines customarily sold in bottles with clearly defined shapes, the following apply:

- Neck levels (i.e. the highest) are normal fills for young wines;
- Bottom neck levels are acceptable for all wines;
- Neck/top shoulder levels are acceptable for most wines;
- High shoulder levels are normal for bottles 15 – 20 years old;
- High to mid-shoulder levels are acceptable for any wine over 20 years old;
- Mid-shoulder levels may carry some risk or weakening of the cork;
- Low shoulder levels are quite risky;
- Bottom shoulder levels are only for rare or collectable wines which may be undrinkable.

In bottles without pronounced shoulders (i.e. with gently sloping sides), such as for Burgundy, German and Alsatian wines, the distance from the wine level to the underside of the cork is measured and quoted in inches. If the estate or merchant does not disclose any distance, then it must be less than 1inch.

The level of wine in the bottle is one of its necessary specifications and criteria of appearance. *Cheers!*

Appendix XVII

The Metric Definitions of Imperial Units

Imperial units nowadays tend to be defined in terms of the corresponding metric ones. In practice, they are the same as they have always been, so that the historical continuity is not broken. The new definitions do have two distinct advantages: (a) the correspondence between the metric and imperial measures is exact; and (b) if the physicists come up with something clever that determines any of the metric units more precisely, then that increased precision is automatically reflected in the imperial system.

When Concorde was designed, the French engineers worked in metric units and the British in imperial. Each set of calculations served as a check on the other, and there were no mistakes. Here are the main definitions.

> 1 yard = 0.9144 metres
> 1 pound (avoirdupois) = 0.45359237 kilograms
> 1 gallon = 4.54609 litres

Armed with only that much information, it is possible to calculate equivalents for most other British units. For example, the British Thermal Unit was the heat required to raise the temperature of 1 pound of water by 1 degree Fahrenheit. In the same way, the calorie was the heat required to raise the temperature of 1 gram of water by 1 degree Celsius. Since 1F degree = $5/9^{ths}$ of 1C degree and 1 pound = 453.59237 grams, therefore 1BTU = $5/9^{ths}$ of 453.59237 or very nearly 251.9957611 calories.

It was established early in the 19th century that heat is a form of energy, and experiments were done to determine the quantative relationship. Of course, the answer depends on the conditions in which the experiments are carried out, but the calorie always come to about 4.18 or 4.19 joules. It was finally decided that the calorie should be redefined as 4.1868 joules. This being so, the BTU comes to exactly ($^{5}/_{9}^{\text{ths}}$) (453.59237) (4.1868) or 1,055.05585262 joules.

Likewise, the kilogram force was defined to be exactly 9.80665 newtons. It was by using this fact and the definitions of the foot and the pound that we were able to calculate the horse-power as being exactly 745.69987158227022 watts.

Appendix XVIII

Coherent Systems of Units

If we are using the foot-pound-second system of units, then we have to measure speed in feet per second and not, say, in miles per hour. Similarly, the unit of force has to be the force which gives a particle of mass 1 pound an acceleration of 1 foot per second per second (the poundal). A system of units constructed by this principle is said to be *coherent*.

Working within a coherent system greatly simplifies the equations used by physicists. Indeed, many books on theoretical physics don't specify a system of units at all – it is just assumed that a coherent system is being used. For example, Einstein's famous equation $E = mc^2$ works just as well with the energy in joules, the mass in kilograms and the speed of light in metres per second. For ordinary mechanics, the foot-pound-second system is no better and no worse than the metre-kilogram second system.

If expressed in imperial terms, the speed of light being 186,000 miles per second, then a pound of mass possesses an amount of energy equivalent to a million-ton meteor travelling at about 20,000mph. Whereas, using metric units, the speed of light being 300,000 kilometres a second, then that same meteor (weighing 1,016,046,900kg or 1,016,047 metric tonnes) would be travelling at about 32,000kmh – which exactly corresponds to 20,000mph! So what's the problem?

The fact that the imperial system does not include electrical units does not imply that British scientists were behind their continental colleagues in discovering the properties of electricity. In his textbook on Electrodynamics, the German physicist Arnold Sommerfeld gives biographical notes on four eminences in this field: the Englishman Michael Faraday, the Scot James Clerk Maxwell, the Frenchman André Marie

Ampère and the German Heinrich Hertz. Faraday discovered, among other things, the principle of the electric dynamo. Without this invention, we should have no electricity in our homes. However, Maxwell's equations are perhaps the most beautiful in physics; prompting Boltzmann to quote "Was it God who wrote these lines...". The equations do not specify a set of units; for it is merely assumed that a coherent system is used. Maxwell's theory lead naturally to Einstein's theory of relativity.

The metric system is logical, dogmatic, comprehensive and coherent (in the sense described above). These characteristics make it suitable for most scientific work, particularly physics. The British system is practical, flexible and contains a choice of coherent sub-systems. These characteristics make it suitable for much scientific work and for virtually all everyday applications.

When we considered Lewis Carroll's gravity train [see Appendix VIII], we worked in the foot-pound-second system and only converted the time to minutes at the end of our calculations. The foot-pound-second system is a coherent sub-system of British units and, until the 1960s, was the one usually used for secondary-school mechanics [see, for example, D. Humphrey: Intermediate Mechanics].

In Lewis Carroll's story, Lady Muriel says: "But the velocity in the *middle* of the tunnel must be something *fearful!* Since distances for train journeys are normally measured in miles and speeds in mph, the principle of coherence requires that times are measured in hours rather than in minutes. So the journey time of 42 minutes becomes $^7/_{10}{}^{ths}$ of an hour. Therefore, if the length of journey is 350 miles, then the average speed is $350 \times {}^{10}/_7 = 500$mph. But the speed in the middle of the tunnel is $250 \times pi$ [3.1415926] which equals very nearly 785.4mph. Bear in mind that the speed of sound is about 750mph!

In any coherent system, there cannot be more than one unit for each type of magnitude. Indeed, it is clear that if there were (say) two different units of weight, then one of them, at least, would not fit coherently into the system. Here is a short list of some commonly used units with their metric equivalents.

kilowatt-hour	3.6 megajoules
ampère-hour	3.6 kilocoulombs
calorie	4.1868 joules
millibar	100 pascals

As the joule, the coulomb and the pascal are all metric units, and as the metric system is coherent, it follows that the units in the first column are not metric and the British people are being hoodwinked by those in authority who pretend that they are. Thus, the Department of Trade and Industry has confirmed that the kilowatt-hour is not a metric unit yet British Gas tells its customers that it is – having abandoned the therm in order to 'go metric' but replacing it, not with the metric joule, but with the hybrid kilowatt hour. Likewise, when energetic values are given on food-packets, it is mandatory to give them in the non-metric calorie as well as in the metric joule.

These examples show that the official policy of so-called metrication is in fact less concerned with imposing the use of proper metric units than merely with banning the use of our customary units. As a consequence it seems that we are in danger of ending up with a collection of units that has neither the theoretical advantages of the metric system nor the practical advantages of the British system.

Not long ago, when the United Kingdom was still a free, democratic sovereign state, we were able to choose which units we used. To ban traditional units now, after they have been in use for thousands of years, is rather like the Greek colonels of the 1960s banning the plays of Sophocles. One of the aims of this little book is to make the case for a return to that freedom of choice.

In 1340, the weight of a sack of wool was standardized as 364 lbs, i.e. 26 stones. That provided 1lb for each day of the year bar Christmas Day, a clove (7lb) for every week, a stone per fortnight or a tod (28lbs) per lunar month.

Appendix XIX

Hall-Marking

*not directly concerned with the imperial system but part of
the same culture and trading traditions – and equally on the
defensive against the envious vandals of the EU*

from *The Times* of 2 October 2003

"When Edward I passed a royal statute introducing hall-marking in 1300, he established one of the oldest systems of consumer protection in the world. Frustrated by people debasing gold and silver with cheaper metals, he made it law that anything made of either precious metal had to be independently tested and stamped with an official mark before it was sold.

Seven hundred years later, the system remains virtually unchanged. Jewellers had to send their wares to an 'assay' office, which would test the quality of the metal and stamp it with a series of three marks showing who made it, the purity of the metal, and the stamp of the assay office.

The oldest assay mark is the leopard's head of the London Assay Office, which has been in continuous use since the wardens of the Goldsmiths' company were given responsibility for marking gold and silver wares with the King's mark of the leopard's head. But silversmiths and goldsmiths resented sending their wares all the way to London; and over succeeding centuries other assay offices were set up around Britain. Today there are four still in existence. The other three are: Edinburgh, whose symbol is a castle; Birmingham, which uses an anchor, and Sheffield, which has a crown.

Birmingham and Sheffield are thought to have got their symbols from a 'Crown and Anchor' pub where negotiations were held to set up the two different offices. The hall-marking act of 1973 extended hall-marking to platinum, and standardised the symbols used. In 1997, laser marking was introduced for hollow articles such as watch-cases that could be damaged by stamping. In 1999, to standardise trade across Europe, the traditional marks such as the Lion Passant for 92.5% pure silver became optional: instead, fineness had to be expressed by a number showing parts per thousand."

But now, as *The Times* reported on that date, the European Commission planned, by means of its Precious Metals Directive, to compel the UK to abolish this hall-marking system that for over seven hundred years has guaranteed the quality of jewelry in Britain and our exports world-wide. Instead, jewellers would be allowed to certify the quality of their own products and stamp it with their own EU standard 'hall-mark'.

(Five other EU member states also have statutory hall-marking, while the rest use a variety of different systems.) The UK government was bitterly opposed to the directive, as were consumer groups and jewellry manufacturers; but the directive came under 'qualified majority voting' which meant that it could be forced through if not enough countries support Britain.

The USA has no independent checks on jewellers, and a recent study of jewellry sold in New Jersey found that two thirds was 'under-carat'. Australia, which went from statutory independent hall-marking to self-marking found that the percentage of under-carat goods rose from 1% to 15%. To quote a British Jewellers' Association spokesman: "The record of trading standards officers in testing the standards of weights and measures is pretty poor. If they can't cope with weights and measures, which are really simple, how can they cope with assessing the fineness of precious metals?"

Italy, with the largest jewelry industry in Europe, was promoting the directive, which would give it greater access to Britain, the biggest market in Europe. With no statutory hall-marking, Italy has the biggest jewelry *black* market in Europe! She was exploiting her occupancy of the Presidency of Europe for the latter half of 2003 to set the agenda for new legislation. As the first such proposal was defeated in 1993 and the second in 2003, it will doubtless be revived again and again until it finally gets through and then can never be reversed – as with referenda on EU membership or adoption of the euro. The barbarians are always at the gate and may eventually break it down. So it is all the more vital now to record and cherish our ancient knowledge and traditions, whether of weights and measures or hall-marking or even democracy itself. Ultimately, after this period of oppression, they may emerge and flourish more strongly than ever.

A Table spoon of 1785. For silver there are two standards, Sterling (92.5%) and Britannia (95.8%). Sterling silver in England is represented within a hallmark by a lion passant and Britannia standard by a seated figure of Britannia holding a spear and shield. A lion rampant or thistle represents the Sterling standard in Scotland and a harp crowned in Ireland. On the right is a rare duty mark (originally struck incuse as opposed to the other marks which are struck intaglio [in cameo form]).

Appendix XX

A Short History of the Printer's Point

From the 1450s, when Gutenberg started printing with movable type, there was a need to identify and measure the type used. But for almost three centuries, type sizes were differentiated only by name and bore no relationship to each other – nor did the sizes of one type founder correspond with another.

It was Father Sébastien Truchet who first devised a system for measuring type. In 1694 he proposed that the 'line' (a Silversmiths' unit measuring $\frac{1}{12}$th of a Royal French inch) could be divided into 12 'points': these would provide the numerical increments for a range of type sizes. The scheme failed to take off but was revisited in 1737 when Pierre Simon Fournier proposed a point based on $\frac{1}{12}$th of a 'Cicero' (a French type size measuring 0.1648"). The Fournier point gained favour with many parts of the industry but lacked a reliable reference. This was overcome in 1783 when type founder François-Ambroise Didot redefined the point using almost the same progressions as Truchet. The result was the Didot point which measured $\frac{1}{72}$nd of a Royal French inch. The Didot point (0.014775") became the standard in Europe, even after the declaration of the metric point (0.4mm) in 1790.

Despite the consensus in continental Europe, British and American type foundries were unable to agree a standard and continued to produce their own sizes, each identified by name rather than measurement. Change finally came in 1871 when the molds of a leading Chicago type foundry were destroyed by fire. The company, Marder Luse, were forced to start afresh and as a bench mark adopted the 'Johnson Pica' (the Anglo-Saxon equivalent to the Cicero) which was already produced by several American foundries. Marder Luse's manager in San Francisco was inventor Nelson C Hawks who had been been working on a new type system for some years. Like those before him in France,

Hawks saw that duodecimal principles were ideal for type and divided the Pica into 12 points, making its half-size 'Nonpareil' 6 points. The 'Johnson Pica' had its origins in equipment supplied by Fournier in the 1700s, and the result was a point size of 0.01383 inch – very close to Fournier's 0.0137 inch. The new system allowed type to be produced in one point increments, and most of the old irregular sizes were abandoned along with their names (see table opposite). In 1886 Hawks persuaded the American Typefounder's Association (ATA) to adopt his system as a national standard but, in agreeing to do so, they decided that the new point should have a metric equivalent, as follows:

83 picas (996 points) = exactly 35cm
1 pica (12 points) = approx 4.217mm
1 point = approx 0.3514mm (approx 0.01383 inch)

In practice however, the industry usually quoted 0.01383 inch when defining a point – approximately $1/72^{nd}$ inch. The United Kingdom adopted the system in 1898.

There was also a system of word spaces, which applied regardless of type size. The examples below are 10pt.

▮	em (mutton)	=	the square of the body
▯	en (nut)	=	$1/2$ body or 2 to the em
▯	thick space	=	$1/3$ body or 3 to the em
▯	middle space	=	$1/4$ body or 4 to the em
▯	thin space	=	$1/5$ body or 5 to the em
│	hair space	=	$1/12$ body approx

The 12pt em (Pica) was treated as a unit in its own right and used for measuring the lengths of lines of type. For example, this paragraph is set to a width of 23 ems. Another old name to survive was the 'Agate', which was originally the very small type used in the classified sections of newspapers (approx 5.2pt). Whilst Agate sized type has not been cast for over a century, American newspapers still use the Agate scale to measure the depth of a column of type. 14 agates = one (column) inch.

The metal type on which the point system was based is now largely a thing of the past, but the point has been adapted to computer use. When Adobe launched their page description language, the Postscript point was defined as being precisely $1/72^{nd}$ inch (0.01388 inch). The $1/72^{nd}$ inch definition – approximated since 1886 – was finally made exact! However, in the tradition of all that came before, an alternative typesetting language designed by TeX fixed the point at $1/72.27$ inch – almost identical to the ATA point.

OLD TYPE NAMES AND THEIR APPROXIMATE SIZES

Excelsior / Minikin	3 pt
Brilliant	4 pt
Diamond	4.2 pt
Pearl	4.8 pt
Ruby / Agate	5.2 pt
Nonpareil	6 pt
Emerald / Minionette	6.8 pt
Brevier	7.8 pt
Bourgeois	8.5 pt
Long Primer	9.6 pt
Small Pica	10.5 pt
Pica	12 pt
English	13.5 pt
2-Line Brevier / Columbian	15 pt
Great Primer	16.9 pt
Paragon	19.4 pt
Double Pica	20.8 pt
2-line Pica	24 pt
2-line English	27 pt
Four line Brevier	32 pt
2-line Great Primer	33.9 pt
Canon	48 pt

Appendix XXI

Threads in Engineering

Sir Joseph Whitworth was the first to propose a standardised thread form in 1841. The principal features of the British Standard Whitworth (BSW) thread form are that the angle between the thread flanks is 55 degrees and the thread has radii at both the roots and the crests of the thread.

The British Standard Fine (BSF) thread has the same profile as the BSW thread form but was used when a finer pitch was required for a given diameter.

British Association (BA) was used for small diameter threads (below $1/4"$ diameter). The thread has radiused roots and crests and has a flank angle of 47.5 degrees. The sizes are based on a number sequence, specifying the size and the pitch. There is no coarse or fine. The lower the number, the larger the bolt size, so 0 BA is the largest, and 23 BA the smallest.

Whitworth was also adapted for pipe threads. British Standard Pipe (BSP) threads were 55 degree form but were nominated by the bore of the pipe. i.e. $1/4"$ BSP was about .528"O/D and 20TPI.

There are other thread systems, such as B S Cycle; these have Whitworth 55 degree forms but nearly all have a constant 26 TPI except the very small.

The U.S. equivalents are called the Unified Thread System, formally known as the American Standard but renamed in 1949. These have 60 degree thread flanks.

• SAE UNC or Unified National Coarse is the counterpart to BSW

• SAE UNF Unified National Fine is equivalent to BSF.

• The SAE number series are the equivalent to the BA series

UNF and UNC spanner markings correspond to the dimension of the hexagonal bolt head measured across the flats. Whitworth spanner sizes indicate the diameter of the bolt itself.

BRITISH STANDARD WHITWORTH (BSW)

Size	O/dia	TPI	Equiv UNC TPI
1/8	.125	40	
3/16	.1875	24	
1/4	.25	20	20
5/16	.3125	18	18
3/8	.375	16	16
7/16	.4375	14	14
1/2	.500	12	13
9/16	.5625	12	12
5/8	.6250	11	11
3/4	.7500	10	10
7/8	.8750	9	9
1	1.000	8	8

BRITISH STANDARD FINE (BSF)

Size	O/dia	TPI	Equiv UNC-TPI
3/16	.1875	32	
1/4	.25	26	28
5/16	.3125	22	24
3/8	.375	20	24
7/16	.4375	18	20
1/2	.500	16	20
9/16	.5625	16	18
5/8	.6250	14	18
3/4	.7500	12	16
7/8	.8750	11	14
1	1.000	10	12

BRITISH ASSOCIATION (BA)

Size	O/dia	TPI
0	.2362	25.4
1	.2087	28.2
2	.185	31.4
3	.1614	34.8
4	.1417	38.5
5	.1260	43.1
6	.1102	47.9
7	.0984	52.9
8	.0866	59.1

The decline in the use of Whitworth threads started during WWII when it became apparent that a common thread system was desperately required for the Allies (or more strictly, those countries using the inch in engineering). In 1948, the British chose to embrace the U.S. SAE thread system but this transition took many years to complete. In 1965 the British Standards Institution issued a policy statement requesting that organisations should regard the BSW, BSF and BA threads as obsolescent. The first choice replacement for future designs was to be the ISO metric thread with the ISO inch (Unified) thread being the second choice.

SELECT BIBLIOGRAPHY

Units and Standards of Measurement
Dept. of Scientific and Industrial Research, London, 1951

Our Weights and Measures
H J Chaney: Eyre and Spottiswoode, 1897

Tables of Physical and Chemical Constants
(16th edition) Kaye and Laby: Longman 1995
[this remains the 'Bible']

For Good Measure
William D Johnstone: NTC Publishing Group, Chicago, 1998
ISBN 0 8442 0851 5

The Sizesaurus
Stephen Strauss: Kodansha America Inc, New York, 1995
ISBN 1 56836 110 6

How Heavy, How Much and How Long?
Colin R Chapman: Lochin Publishing, Dursley, GL11 5RS,
1995 ISBN 1 873686 09 9

Ancient Weights and Money
J H Parker: Oxford, 1836

The Law of Weights and Measures
J A O'Keefe: Butterworth, 1966/1978

Measuring America
Andro Linklater: Harper Collins, 2002
ISBN 0 007 10887.7

Weights and Measures
J T Graham (revised by M Stevenson): Shire Publications,
Princes Risborough, 1987

A Beginner's Guide to Stone Circles
Robin Heath: Hodder and Stoughton, 1999
ISBN 0 340 73772 7

190

The Dimensions of Paradise
John Michell: Thames and Hudson, 1988
ISBN 0 500 01386 1
and *Ancient Metrology* John Michell, 1981

The Ascent of Man
Jacob Bronowski: BBC, 1973
ISBN 0 563 104988

Opus 2 – All Done with Mirrors
John Neal 2000
ISBN 0-9539000-0-2   johnneal@secretacademy.com

Apollo in Albion (working title)
Anne Macaulay: (pending) 2006

Weights and Measures, origins and development in Great Britain up to 1855
F G Skinner: HMSO, 1967

The Weights and Measures of England
R D Connor: HMSO, 1987
ISBN 0-11-290435-1

Dictionary of English Weights and Measures
R E Zupko: University of Wisconsin Press, 1968

Historical Metrology
A E Berriman: Dent, 1953

The Story of our Weights and Measures
Edward Nicholson: 1901

Men and Measures
Edward Nicholson:1911
www.metrum.org/measures/whystud.htm&/romegfoot.htm
L C Stecchini

How Many? A Dictionary of Units of Measurement (revised 2001),
R Rowlett: University of North Carolina (Chapel Hill)

The Great Gram Scam
V T Linacre: BWMA, 2002

THE BRITISH WEIGHTS AND MEASURES ASSOCIATION

Patrons: Lord Monson, The Hon Mrs Gwyneth Dunwoody, Sir Patrick Moore and Vice Admiral Sir Louis Le Bailly.

Honorary Members: Peter Aliss, Clive Anderson, Trevor Bailey, Michael Barry, Christopher Booker, Ian Botham, Max Bygraves, Beryl Cook, Jilly Cooper, Professor Richard Demarco, Roy Faiers, Sir Ranulph Fiennes, Edward Fox, Dick Francis, George Macdonald Fraser, Sandy Gall, Candida Lycett Green, Simon Heffer, Peter Hitchens, Jools Holland, Professor Richard Holmes, Conn Iggulden, Hal Iggulden, Richard Ingrams, Dr James Le Fanu, Jonathan Lynn, Alexander McCall-Smith, Dr Richard Mabey, Christopher Martin-Jenkins, Robin Page, Lord Phillips of Sudbury, RWF Poole, Sir Tim Rice, Andrew Roberts, JK Rowling, Quinlan Terry, Keith Waterhouse, Sir Rowland Whitehead, Anthony Worrall Thompson.

BWMA gratefully records the Honorary Membership of the late: Lord Shore PC, John Aspinall, Nirad C Chaudhuri CBE, Fred Dibnah, Sir Julian Hodge, Bernard Levin CBE, Leo McKern AO, Norris McWhirter CBE, Jennifer Paterson, David Shepherd MBE, Charles H Sisson CH, Fritz Spiegl, FS Trueman OBE.

BWMA is a non political voluntary association, engaged in metrological research as well as in the campaign to restore freedom of choise between the imperial and metric systems. Office-bearers give their time freely. Membership is open to all. The annual subscription is £10.00. A journal, *The Yardstick*, is published occasionally, as well as a newsletter, *The Footrule*. The AGM (and Conference) is held in the early Summer.

President, Vivian Linacre: Chairman, Michael Plumbe; Director, John Gardner; Press Officer, David Delaney; Hon Treasurer, Lee Consterdine; Editor, Robert Stevens.

INDEX OF SOURCES

Adams, Douglas 137

Adams, John Quincy 119

Alberti, Leon Battista 77

Auden, W H 78

Bemer, R W 153

Bible 74-76

Blair, Eric A (George Orwell) . . .78

Blake, William 81

Bronowski, Jacob 16, 103

Boyer & Merzbach 82

Burrell, Lawrence 140

Carroll, Lewis (see Dodgson, C L)

Chambers, Robert 13

Chapman, Colin R 39

Cicero 77

Coles, Harry 61

Doczi, György 81

Dodgson, C L 78, 137, 178

Doutre, Martin 83, 105

Elliott, Charlotte 78

Flavius, Joseph 77

Frost, Robert 74, 171

Heath, Robin 90

Herodotus 77

Herschel, Sir John 18

Hesiod 76

Kelvin, Lord 78

Linklater, Andro 35

Macaulay, Anne 61, 79-82

Martin-Jenkins, Christopher . . .138

McGreevey, Thomas 154

Michell, John 18, 84, 88-90

Milton, John 77

Morgan A P & Carpenter O . . .166

Napoleon I 78

Nature (1922) 104

Neal, John 4, 104

Nicholson, Edward 102

Patten, Ian B 16

Perrine F A C 166

Plato 77

Shakespeare, William 110

Smyth, Professor C Piazzi 102

Southey, Robert 78

St Isadore of Seville 77

Strange, John F W 100

Strauss, Stephen 9

Turner, Derek 99

Van der Waerden, B L 82

Whillock, Arthur 13, 103

INDEX OF NAMES AND PLACES

Al Khwarismi 147

American Petroleum Institute . . .52

Ampère, André Marie 178

Angkor Wat 7

Aswan 19

Athelstan 7, 127

Babylonian 1

Bach, J S 60

Berriman, A R 85

Big Bang 160

Boltzmann, Ludwig 177

Bouger, Pierre 155

British Standards Institute 103

British W & M Association . . 99, 191

Casey, Robert W 26

Chaney, H J 141

Clemens, Samuel (Mark Twain). . 21

de la Condamine, C M 155

Cromwell, Oliver 59

Danzig (Gdansk) 9

Didot, F-A 183

Dozenal Society 146

Edison, Thomas 30

Edward I 14, 16, 40

Edward III 40

Einstein, Albert 177

Elizabeth I 85

Eratosthenes 19

Euclid .101

Fahrenheit, G D 64

Faraday, Michael 177

Fibonacci (Leonardo of Pisa) . . .79

Fournier, P S 183

Gregory XIII, Pope 26

Gunter, Edmund 35

Hawks, N C 183

Hertz, Heinrich 178

HMS Hood 22, 23

Hadley, John 6

Henry I 14

Henry II 44

Henry VII 51

Henry VIII 40

Hutchins, Thomas 168

Iohango 4

Jefferson, Thomas 10, 171

Joule, J P 67

Kauffman, Bill 99

Leitz, Ernst 30

Mach, Ernst 70

Maskelyne, Nevil 155

Maxwell, James Clerk 177

Mount Everest 162

Mouton, Gabriel 97

Napoleon I 28

Newton, Isaac 5

Norwood, Richard 6

Ordnance Survey 164

Parthenon 14, 85

Petrie, Sir William F 13

Picard, Jean-Luc 6

Pitt-Rivers 15

Poynting, J H 156

Pyramid (Great) 19

Pythagoras 8, 60, 80, 82

Schiehallion 155

Sèvres .97

Shannon, Claude 145

Sommerfeld, Arnold 177

Stecchini, Livio 127

Sumeria 7, 15, 163

Thom, Alexander 79

Thom, A S 79

Thomson, Wm (Lord Kelvin) . . .64

Thrust SSC 100

Tiahuanco 7

Truchet, Sébastien 183

United Nations 100

Vinci, Leonardo Da 81

Watt, James 68

Weights & Measures Act 1897 . .104

White, Gilbert 161

William of Malmesbury 14

GENERAL INDEX

acre 16, 17, 127-128

acre-inch/foot 47

agate .184

ale and corn gallon 118

Anglo-Saxon rod 16-17

Apothecaries' Fluid Measures . . .54

Apothecaries' Weights 45

Astronomical Measures 24

Avoirdupois Weights 40

baker's dozen 30

Balthazar 54

barleycorn 16, 31, 128

barometer 63

barrel (UK/US) 52, 53

barrel (wine) 54

bed sizes 29

Beaufort Scale 172

Beer and Ale 54

Biblical lengths 29

Bibliography 189

binary 1, 145

bits 145

blood pressure 63

boll 48

Book Sizes 57

bottle .54

braccia 14

breakfastcup 53

British Horse Power 68

British Standard Gauge (wire) . . .166

British Standard Fine 187

British Standard Whitworth 187

British Thermal Unit 67

bullets 163

bushel 48, 51, 118

butt 54

cable 21

calorie 67, 179

carat 16, 44

Celsius 64

cental 40, 45

chain 13, 35

chain (nautical) 21

chalder 144

chop (Scots) 143

cicero 183

circles 163

Circular Measures 37

compass points 37

corbyn 54

Cord 169

Cosmic Numerology 83

Cotton measures 30

cubit 13, 17, 19, 29, 32

cup (US) 53

decimal 1

degree 37

dessertspoon 53

diapason 61

Didot (type sizes) 31

digits 14, 29, 32

drachm 43, 45, 55

dram .40

drops (medicine) 55

drops (Scots) 143

duodecimal 1

duodecimo 56
Earth (dimensions) 8, 18, 25-26, 90
Earth (interior) 164
Earth (weight) 155
Egyptian astronomy 17-20
Egyptian cubit 85
Egyptian foot 85
Egyptian sep weight 40
ell 14, 32
equinoxes 25
em (mutton) 184
en (nut) 184
Essential knowledge 208
falls .144
fathom 17, 21, 32
Film 30
finger .29
firkin .54
flagon .54
Florentine ounce 40
Fluid Measures 51
fluid measures (medicine) 55
folio 56
foolscap 57
foot (imperial)... 13, 17, 89, 127-128
Force .65
French Revolution 28
furlong 13, 16, 35, 127-128
furlot .144
Gallon (article) 122
gallon 48, 51, 54
geographical mile 33
gill 51
Golden Mean (*phi*) 80
grade .37
grain 40, 43, 45, 143

grand pianos 29
Greek cubit 85
Greek foot 14, 85
Greek furlong 85
Greek mile 85
Gregorian calendar 26
gross 30
hairsbreadth 31
Hall-marking 180
hand 31, 32
hank 30
hide 47
hogshead 54
horse-power 68
hundred (long/great) 30
hundredweight 40
imperial folio 57
Imperial system 52
inch 13, 17, 31, 32, 63
Jeroboam 54
jug 54
Julian year 26, 157
kilderkin 54
kilopascal 63
King's Girth 127
Kitchen (In the) 53
knot 21
Lawn Tennis 125
ligne 17
light-second/foot/year 24
line 29
linear measures 13
link 17, 35
lippie .144
liquid oz. 53
log-line 21

lunar month/year 25, 92

lunation 90

Le Marin de Commerce 41

Maundy Pennies 44

Medicine Measures 55

Megalithic Measures 70

megaparsec 26

meridian (latitude & longitude) ...23

met 141

Methuselah 54

Metric definitions 175

metric system ....97 *et seq*, 119

micron .... .31

mil.......................31

mile (statute) .. 13, 33, 90, 127-128

mile (nautical) 33

mile (Greek and Roman) 85-88

mile (meridian) 33

millibar 63, 178

minute 37

minute of arc 17, 21

Moon 18

Mountains 161

Music 60

mutch 143

Mystery of 42 137

nail 14, 31

Nautical Measures 21

nautical mile 21, 33

Nebuchadnezzar 54

nonpareil 184

Northern foot 13, 16

octavo....................56

Oil......................153

organs 61

ounce 40, 43, 45

palm 29, 31, 32, 127-128

Paper (folding) 56

Paper (sheet) 58

parsec 26

peck 48, 51, 118, 144

pennyweight 43

phi (Golden Mean) 80-81

pi (3.1415926) 37, 81

pica 184

pint 51, 53, 118

pollices 14

population 160

porter 141

pottle 51

pound 40, 43, 55, 63

poundal 65, 67

Power 68

Printer's Point (article) 183

Printer's type sizes 31

Printing type – old names 185

Proof Spirit 125

puncheon 54

quadrant 37

quart 51, 118

quarter 40

quarto 56

Queen Ann Wine Gallon 52

quintal 45

radian 26, 37

rainfall 159

Rankine 64

reed 29

Rehoboam 54

remen 15

rider 43

rod (pole/perch) 13, 35, 46-47

200

Roman cubit 85
Roman foot 13, 85
Roman furlong 85
rood 46
Salmanazar 54
saltspoon 53
score 1
Scotch .51
Scotland 140
scruple 45
second 25, 37
sexagesimal 1
sextant .37
shackle 22
shaftment 32
shekel 16, 45
shoe sizes 29
Shotgun bores 30
sidereal day/year 25
skein 30
solar month/year 25, 92
span 29, 32
Speed .70
spindle .30
stade 15, 29
stadia 8
Standard Linear Measure 13
steelyard 39
stone 40
Stone Circles . . . . .63, 79 *et seq*
Stress and Pressure 63
strike 48
Sun 158
Surveyors' Measures 35
Système International . . . 28, 67, 98
tablespoon 55
Talmudic measures 29

tankard .54
teacup/spoon 53
Temperature 64
therm .67
thread .30
Threads in Engineering 186
Time 25
tod 179
toise de Perou 17
ton (short/long) 3, 40
ton (pressure) 63
tonnage 23
tonne 41, 53
tonneau 41
Treaty of Union 1707 140-1
trones (scales) 40, 141
tropical year 25
Troy Weights 63
tun 54
tyre pressure 63
United Thread System 186
US measures (differences) 2-3
US fluid measures 52
Volume and Cubic Capacity 48
watt 68
Weight of Water 53
Weight and Mass 39
Weights and Measures Act 1985 .63
Winchester bushel 52
Wine 54
Wine levels 174
Wire Gauges 166
Work and Energy 67
yard (imperial) 13, 17
yard (land) 47
Zereth measures 29

NOTES

NOTES

NOTES

NOTES

NOTES

Notes

NOTES

ESSENTIAL KNOWLEDGE

12 inches = 1 foot
3 feet = 1 yard
22 yards = 1 chain
10 chains = 1 furlong
8 furlongs = 1 mile

16 drams = 1 ounce
16 ounces = 1 pound
14 pounds = 1 stone
2 stones = 1 quarter
4 quarters = 1 hundredweight
20 hundredweight = 1 ton

20 fluid ounces = 1 pint
2 pints = 1 quart
4 quarts = 1 gallon

4,840 square yards = 1 acre
640 acres = 1 square mile

water freezes at 32° Fahrenheit and boils at 212°

normal body temperature is 98.4°

one gallon of water weighs 10 pounds